Higher Mathematics from an Elementary Point of View

H. Rademacher

Edited by D. Goldfeld

Notes by G. Crane

Birkhäuser
Boston • Basel • Stuttgart

H. Rademacher
(1892–1969)

Editor
D. Goldfeld

Library of Congress Cataloging in Publication Data

Rademacher, Hans, 1892–1969.
  Higher mathematics from an elementary point of view.

  Bibliography
  Includes index.
  1. Numbers, Theory of. I. Goldfeld, D. II. Title.
QA241.R214   1982   512'.72   82–14722

CIP – Kurztitelaufnahme der Deutschen Bibliothek

Rademacher, Hans:
Higher mathematics from an elementary point of
view: from lectures given at Stanford Univ.
summer 1947 / H. Rademacher. Ed. by D. Goldfeld.
Notes by G. Crane. — Boston; Basel;
Stuttgart: Birkhäuser, 1982.
  ISBN 3-7643-3064-3

With 58 illustrations.

Printed in the United States of America.

ISBN 3–7643–3064–3

Preface

This book is a lively account of lectures given by H. Rademacher at Stanford University in 1947. It deals with several topics in elementary number theory; namely, prime numbers, Farey fractions, decimal fractions, approximation of irrational numbers by rationals, the exclusion-inclusion principle, Ford circles, the modular group and modular functions, and linkages. The choice of topics is quite unique for this type of book and although the treatment and subject matter is elementary, one cannot come away from a reading of this material without a feeling of profound depth.

The first chapter examines new advances (in 1947) in the theory of prime numbers, to which I have added, in the form of notes on the chapters, the most recent results to date. Rademacher then discusses unique factorization, rational numbers, decimal expansions, continued fractions, Farey series, and approximation of irrationals by rationals. Again, I have tried to update and comment on some of this material in the notes. The multiplicativity of certain number-theoretical functions, such as Euler's ϕ-function, is deduced as a consequence of the exclusion-inclusion principle. The book then goes on to define Ford circles. By use of these circles, an elementary proof of Hurwitz' theorem is obtained. The Ford circles also have a geometric representation as fractions allowing a new treatment of Farey fractions. The discussion of the modular group and modular functions is very lowbrow and gives a unique insight into this important subject by way of several beautiful and elementary examples. In fact, the section on functions on groups can be taken as an elementary introduction to harmonic analysis. There are even hints at the theory of Eisenstein series. Finally, the last chapter, on linkages, provides a fascinating account of a mechanical device which is almost unknown to mathematicians.

I would very much like to thank Birkhäuser for the honor of allowing me to edit this work. My own contributions have been very minimal and I trust that my notes are accurate.

D. Goldfeld
Summer 1982

Contents

Chapter 1

Prime Numbers

Prime numbers *are such integers as have only one and themselves for divisors,* i.e.,

$$2, 3, 5, 7, 11, 13, 17, 19, 23, 29, \text{ etc.}$$

Most of these are odd numbers; in fact only one of them is even. However, this is not a very deep property, for actually only one prime is divisible by three, etc.

The number 1 is not considered as a prime number, for it gives no additional information concerning the nature of a number when it is decomposed into prime factors, e.g., $12 = 2 \cdot 2 \cdot 3$. In such multiplicative building up of numbers we construct (or decompose) a number from (into) prime factors. Of course our numbers can be built up in an additive manner, e.g., $6 = 2 + 4 = 1 + 1 + 1 + 3 = 1 + 1 + 1 + 1 + 1 + 1$. But reducing all numbers to a sum of units is of little interest.

At the very beginning of the list, the primes are quite dense; however, they become less dense as we proceed to higher numbers. This seems reasonable as we expect a high number to have a greater chance of being divisible by a prime number than a low one. Now, do the primes stop altogether? Is there a point in the list of primes after which there are no more? This question is posed and answered in the Elements of Euclid (which contains much of number theory) in the following way.

Euclid asserts that *there can be no last prime number.* Consider

$$2 \cdot 3 \cdot 5 \cdot 7 \cdot 11 \cdot \cdots \cdot p = N,$$

the product of the prime numbers less than or equal to p. Now $N + 1$ is *not* divisible by any member of the sequence of primes 2, 3, 5, 7, . . . , p, for $N + 1$ has a remainder 1 when divided by each of them. Thus, there are two possibilities: either,

1) $N + 1$ is a prime which is not contained in the list 2, 3, 5, ... , p; or
2) $N + 1$ may be decomposed into other prime numbers which are not contained in the list of primes less than p.

Thus in either case it is shown that there must be a prime greater than p. For example:

$$2 \cdot 3 + 1 = 7, \qquad \text{a prime.}$$
$$2 \cdot 3 \cdot 5 + 1 = 31, \qquad \text{a prime.}$$
$$2 \cdot 3 \cdot 5 \cdot 7 + 1 = 211, \qquad \text{a prime.}$$
$$2 \cdot 3 \cdot 5 \cdot 7 \cdot 11 + 1 = 2311, \qquad \text{a prime.}$$
$$2 \cdot 3 \cdot 5 \cdot 7 \cdot 11 \cdot 13 + 1 = 30\,031.$$

If it is not a prime, its prime factors must be above 13; and indeed $30\,031 = 59 \cdot 509$. This proof of Euclid has long been admired as it is not excelled in directness and simplicity.

Can we get more out of this proof? Can we apply it to other problems? It is possible to classify the prime numbers, leaving aside 2, with respect to 4, as a prime is either of the form $4n + 1$ or $4n + 3$.

$$4n + 1: \qquad 5, 13, 17, 29, 37, \ldots ,$$
$$4n + 3: \qquad 3, 7, 11, 19, 23, 31, \ldots .$$

Is one of these sequences of primes an ending sequence, a finite sequence? The answer is no. And we can easily show that $4n + 3$ *is in fact an unending sequence.* Consider the product

$$4 \cdot 3 \cdot 7 \cdot 11 \cdot 19 \cdot 23 \cdot \cdots \cdot q = M,$$

where q is a number of the sort $4k + 3$. M is a four-fold number, so that $M - 1$ is of the form $4k + 3$. As before: $M - 1$ may be a prime, or

$$M - 1 = p_1 p_2 p_3 \cdots p_l,$$

where the primes $p_1, p_2, \ldots , p_l$ are all different from 3, 7, 11, ... , q. The trick is to show that among the sequence of primes, p_i, there must be one prime of the form $4k + 3$, i.e., that a number of the sort $4k + 3$ must be among the prime factors of $M - 1$. But

$$(4m + 1)(4n + 1) = 16mn + 4m + 4n + 1$$
$$= 4(4mn + m + n) + 1 = 4K + 1,$$

so that a product of the form $(4m + 1)(4n + 1)$ must be of the sort $4k + 1$. However, $M - 1$ is of the form $4k + 3$; so that all its prime factors cannot be of the sort $4k + 1$, whence at least one of them is of the form $4k + 3$ and thus the sequence of primes of the sort $4k + 3$ is unending, as we can always find a prime of this form that is different from all members of the sequence 3, 7, 11, 19, 23, ... , q, where $q = 4k + 3$.

Leaving aside the primes 2 and 3, the remainder of the prime numbers can be divided into two classes, $6n + 1$ and $6n + 5$. For the latter we can use the same trick to show that *the sequence of primes of the sort $6n + 5$ is in fact an unending sequence.* We merely sketch the proof which begins: Form

$$6 \cdot 5 \cdot 11 \cdot 17 \cdot 23 \cdot 29 \cdots \cdot r = R,$$

where r is of the sort $6n + 5$. $R - 1$ is not divisible by 6, 5, 11, ... , r; but R is a six-fold number, so that $R - 1$ has the remainder 5 when divided by 6. Thus as before there must be a prime of the sort $6n + 5$ which is greater than r. Unfortunately this trick is not applicable to the sequences of the forms $4n + 1$ and $6n + 1$.

Let us put a more general question. The numbers of the form $4n + 1$ —

$$1, 5, 9, 13, 21, 25, 29, 33, \ldots -$$

whether they are prime or not, form an arithmetic progression. In general we can write such a sequence in the form $d \cdot n + r, n = 0, 1, 2, \ldots$, yielding

$$r, r + d, r + 2d, r + 3d, r + 4d, \ldots .$$

It would be hopeless to look for prime numbers in such a sequence if r and d had a common factor. So we assume that r and d are coprime. For example:

$$d = 10, r = 3, \text{ yields } 3, 13, 23, 33, 43, 53, \ldots ,$$

where the difference, d, has no common factor with the first term, r. If d and r have no common divisor, does this sequence contain infinitely many primes? G. P. Lejeune Dirichlet (1805–1860) proved this in 1837.

But first let us consider a proof due to Euler of Euclid's proposition that *the number of primes cannot be finite.* Let s be a number greater than 1. Then

$$\frac{1}{1^s} + \frac{1}{2^s} + \frac{1}{3^s} + \frac{1}{4^s} + \cdots = \frac{1}{1 - 1/2^s} \cdot \frac{1}{1 - 1/3^s} \cdot \frac{1}{1 - 1/5^s} \cdots$$

is convergent for $s > 1$. This we can write more briefly as

$$\sum_{n=1}^{\infty} \frac{1}{n^s} = \prod_{p}^{\infty} \frac{1}{1 - 1/p^s},$$

where p is a prime.

How does Euler's identity arise? We know

$$\frac{1}{1 - a} = 1 + a + a^2 + a^3 + \cdots + a^n + \cdots$$

Using this to expand the right-hand side we have:

$$\prod_{p}^{\infty} \frac{1}{1 - 1/p^s} = \left(1 + \frac{1}{2^s} + \frac{1}{2^{2s}} + \frac{1}{2^{3s}} + \cdots\right) \cdot \left(1 + \frac{1}{3^s} + \frac{1}{3^{2s}} + \frac{1}{3^{3s}} + \cdots\right)$$
$$\cdot \left(1 + \frac{1}{5^s} + \frac{1}{5^{2s}} + \frac{1}{5^{3s}} + \cdots\right) \cdots .$$

And from the uniqueness of the factorization of a number into prime factors, it follows that any term on the left will be formed once and only once when this product is multiplied out. Thus, this is a formal identity. Now if $s \to 1$, we know that $\Sigma_1^\infty 1/n$ is divergent. But the identity must hold for all s — so there must be infinitely many factors on the right, if the identity is to hold.

$\Sigma_1^\infty 1/n^s$ depends only on s and when s is any complex number $\zeta(s) = \Sigma_1^\infty 1/n^s$, a meromorphic function with $s = 1$ as a pole, is known as the *Riemann ζ-function*. It is of considerable importance in number theory and is so named after Riemann (1826–1866), who discussed it in a famous paper published in 1860.

Euler's proof tacitly assumes that any number can be decomposed into prime factors and that this decomposition can be accomplished in just one way. This must be proved, but this we shall defer until the next section. A more serious objection is that we have used the notions of infinite series and products and the notion of a limit, none of which are elementary. Now let us seek a proof modeled on Euler's — one that is truly elementary.

To eliminate these notions, let us put $s = 1$ and replace the infinite series and product of Euler's identity by a finite sum and product. Since Euler's identity no longer holds, we must seek a new relation:

$$\sum_{n=1}^{\nu} \frac{1}{n} \overset{?}{=} \prod_{p=2}^{P} \frac{1}{1 - 1/p},$$

where ν and P are numbers to be chosen.

First let us choose a number N, preferably a large one, and form the sum $1 + \frac{1}{2} + \frac{1}{3} + \frac{1}{4} + \cdots + 1/(2^N - 1) + 1/2^N$, i.e., choose $\nu = 2^N$. And then let P be the largest prime less than 2^N, which is surely composite, so that we consider the sequence of primes, $2, 3, 5, 7, \ldots, P < 2^N$.

If we make N larger, and do so continuously, then we expect that the number of primes less than 2^N will indeed increase, for if the set of all prime numbers were finite, then we could not increase the number of primes less than 2^N by taking N larger and larger. This gives us the motivation for a proof: can we show that some strictly increasing function of N is always less than some function of the primes less than 2^N?

Let us write

$$\prod_f = \left(1 + \frac{1}{2} + \frac{1}{2^2} + \frac{1}{2^3} + \cdots + \frac{1}{2^N}\right) \cdot \left(1 + \frac{1}{3} + \frac{1}{3^2} + \cdots + \frac{1}{3^N}\right)$$
$$\cdot \left(1 + \frac{1}{5} + \frac{1}{5^2} + \cdots + \frac{1}{5^N}\right) \cdot \left(1 + \frac{1}{7} + \frac{1}{7^2} + \cdots + \frac{1}{7^N}\right) \cdots$$
$$\cdot \left(1 + \frac{1}{P} + \frac{1}{P^2} + \cdots + \frac{1}{P^N}\right).$$

When $\prod_f$, this finite product, is multiplied out, not only will it contain all of the terms of the harmonic series which precede $1/(2^N + 1)$, but it will also contain some terms which appear later, e.g.,

$$\frac{1}{2^N} \cdot \frac{1}{3^N} \cdot \frac{1}{5^N} \cdot \frac{1}{7^N} \cdot \ \cdots \ \cdot \frac{1}{P^N}.$$

Thus,

$$\sum_{n=1}^{2^N} \frac{1}{n} < \prod_f.$$

But

$$1 + \frac{1}{P} + \frac{1}{P^2} + \cdots + \frac{1}{P^N} < \frac{1}{1 - 1/P},$$

and hence we see that

$$\prod_f < \prod_{p=2}^{P} \frac{1}{1 - 1/p} = \frac{1}{1 - 1/2} \cdot \frac{1}{1 - 1/3} \cdot \frac{1}{1 - 1/5} \cdot \ \cdots \ \cdot \frac{1}{1 - 1/P}.$$

Moreover we can write our partial sum of the harmonic series as

$$1 + \frac{1}{2} + \frac{1}{3} + \cdots + \frac{1}{2^N}$$

$$= 1 + \frac{1}{2} + \left(\frac{1}{3} + \frac{1}{4}\right) + \left(\frac{1}{5} + \frac{1}{6} + \frac{1}{7} + \frac{1}{8}\right) + \cdots + \left(\frac{1}{2^{N-1} + 1} + \cdots + \frac{1}{2^N}\right)$$

$$> 1 + \frac{1}{2} + \left(\frac{1}{4} + \frac{1}{4}\right) + \left(\frac{1}{8} + \frac{1}{8} + \frac{1}{8} + \frac{1}{8}\right) + \cdots + \left(\frac{1}{2^N} + \cdots + \frac{1}{2^N}\right)$$

$$= 1 + \frac{N}{2},$$

where we can neglect the 1 in comparison with $N/2$. Whence we have

$$\frac{N}{2} < 1 + \frac{N}{2} < 1 + \frac{1}{2} + \frac{1}{3} + \cdots + \frac{1}{2^N} = \sum_{1}^{2N} \frac{1}{n} < \prod_f$$

$$< \frac{1}{1 - 1/2} \cdot \frac{1}{1 - 1/3} \cdot \frac{1}{1 - 1/5} \cdot \ \cdots \ \cdot \frac{1}{1 - 1/P}.$$

But

$$\frac{1}{1 - 1/2} \cdot \frac{1}{1 - 1/3} \cdot \frac{1}{1 - 1/5} \cdot \ \cdots \ \cdot \frac{1}{1 - 1/P} = \frac{2}{1} \cdot \frac{3}{2} \cdot \frac{5}{4} \cdot \ \cdots \ \cdot \frac{P}{P - 1},$$

so that we have

$$\frac{N}{2} < \frac{2}{1} \cdot \frac{3}{2} \cdot \frac{5}{4} \cdot \frac{7}{6} \cdot \ \cdots \ \cdot \frac{P}{P - 1}.$$

Now if there were a last prime P, then the right-hand side could have only a finite number of factors and the value would be definitely finite. But this is impossible as we can choose N as large as we like.

Euler's proof of the infinity of primes provides the motivation for Dirichlet's proof of the theorem that *in every arithmetic progression there is an infinity of primes*. This last theorem is much too difficult to consider here.

Up to now everything has been proved before our eyes. The remainder of this chapter will be the assortment of facts about primes, which will be partly proved and partly told.

A list of primes less than 10 000 000 has been computed by D. N. Lehmer (1956). How are such lists prepared? The essential idea seems to have been developed in Alexandria in the Third Century B.C., for it is attributed to Eratosthenes (ca., 250 B.C.) and does not appear in Euclid. Let us write down the list of integers:

1, 2̸, 3̸, 4̸, 5, 6̸, 7, 8̸, 9̸, 1̸0̸, 11, 1̸2̸, 13, 1̸4̸, 1̸5̸, 1̸6̸, 17, ...

First we cross out 2 and all its multiples. The first number, always excluding 1, which is not hit is a prime, for it has no lower divisor. This is 3. So we strike out 3 and its multiples. The next uncancelled number, 5, is prime, having no smaller factor. Eliminating the multiples of 5, we find that 7 is a prime, etc. This purely mechanical method is known as the *Sieve of Eratosthenes*. It depends not upon the numbers themselves, but rather on their position in the sequence of the integers.

We may think of the row of equally spaced dots in Plate I as being continued indefinitely to the right. These we can number, calling the first one 0, the next, to the right, 1, the next 2, etc. The method sketched above is applicable to these dots.

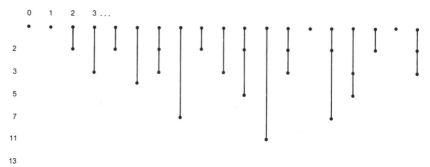

Plate I

Let us cross out every other dot, starting with 0 — a mechanical process. (In Plate I, to avoid confusion this has been done in a second row and lines have been drawn between corresponding dots to indicate the cancellation.) Next we strike out the dots whose distance from the origin is a multiple of the distance of the first dot, excluding always 0 and 1, not previously cancelled. This yields line 3, and repeating the process gives line 5. Stepwise we eliminate all dots that are at multiples of the distance of some previous dot from the origin and find at the end of each step that the first uncancelled dot corresponds to a prime number. Thus it is clear that the Sieve of Eratosthenes depends upon the position of the integers in sequence rather than on the properties of the numbers themselves. This method was the basis on which Lehmer's list of primes was computed.

As n becomes larger, the density of primes decreases. Can this be measured? At fifteen Gauss (1777–1855) asked this question and made a remarkable conjecture. Let $\Pi(x)$ be the number of primes not exceeding x, i.e.,

$$\Pi(1) = 0,\ \Pi(2) = 1,\ \Pi(3) = 2,\ \Pi(4) = 2,\ \Pi(5) = 3,\ \Pi(6) = 3,$$

$$\Pi(7) = 4,\ \Pi(8) = 4,\ \Pi(9) = 4,\ \Pi(10) = 4,\ \Pi(11) = 5,\ \text{etc.}$$

$\Pi(x)$ is a function that increases irregularly — increases by 1 at each prime number and then is constant until the next prime. Gauss conjectured that

$$\Pi(x) \sim \frac{x}{\log x}, \qquad \text{i.e., } \lim_{x \to \infty} \frac{\Pi(x)}{x/\log x} = 1.$$

Although this ratio approaches 1 as x becomes large, the difference $\Pi(x) - (x/\log x)$ increases strongly with x. This conjecture was made in 1792, but it was not proved until 1896 when both Hadamard and de la Vallée Poussin made the proof, based on notions developed by the former in his thesis in 1892, which permitted further properties of the Riemann ζ-function to be deduced. (See G. Birkhoff 1973.) The pressing problem on the ζ-function, which is known as the Riemann hypothesis, is the question of its zeros. It is conjectured that all its zeros, except those on the negative real axis, lie on the line $R(s) = \frac{1}{2}$.

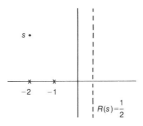

Dirichlet proved that each simple arithmetic progression whose first term and difference are coprime contains an infinite number of primes. Now does the arithmetic progression of second order,

$$2^2 + 1,\ 4^2 + 1,\ 6^2 + 1,\ 8^2 + 1,\ 10^2 + 1,\ \cdots,\ (2x)^2 + 1,\ \cdots,$$

contain an infinite number of primes? No one knows. The answer seems to be beyond our present strength.

Another famous unsolved problem is: Are there an infinite number of twin primes, i.e., primes that differ by 2, e.g., 5 and 7, or 11 and 13? No one knows. Such twin primes are scarcer than prime numbers. The best result in this direction was obtained by a Norwegian mathematician, Viggo Brun (see E. Landau 1958). Curiously the fact that he was isolated from other mathematicians seems to have favored his work — for they, no doubt, would have convinced him that it would be useless to employ the Sieve of Eratosthenes to prove propositions on prime numbers. Using this outmoded device, he was able to show that *the sum of the reciprocals of the twin primes converges,* i.e., $\frac{1}{3} + \frac{1}{5} + \frac{1}{7} + \frac{1}{11} + \frac{1}{13} + \frac{1}{17} + \frac{1}{19} + \cdots$ actually coverges. This is surprising, for it can be shown that *the sum of the reciprocals of all primes diverges.*[1] Thus Viggo Brun's result is an advance. If the number of twin primes is finite, then the sum of their reciprocals certainly converges. While if their number is infinite, it shows that we lose very many primes in extracting this convergent series from the divergent sum of the reciprocals of all primes, so that the twin primes are indeed scarce.

In addition Brun considered numbers that were not prime but consisted of at most 9 prime factors and succeeded in showing that among such numbers there are infinitely many twins, i.e., numbers differing by 2.

Since we know that the primes become less dense as we go to higher numbers and suspect that they appear again and again as twins, we expect that the prime numbers are very irregularly distributed. *Are there arbitrarily large gaps in the sequence of primes? Yes.* For example, consider $M = 1000$: The numbers $M + 2, M + 3, M + 4, \ldots, M + 1000$ are not prime, for

$$
\begin{array}{lll}
M + 2 & \text{is divisible by} & 2, \\
M + 3 & \text{is divisible by} & 3, \\
\vdots & \vdots & \vdots \\
M + 1000 & \text{is divisible by} & 1000.
\end{array}
$$

Thus we construct 999 consecutive numbers no one of which is prime. The same result could be obtained from the product, N, of all the primes less than 1000. Adding 2, 3, 4, 5, ..., 1000 successively to N, we would have again 999 consecutive non-prime numbers, e.g., $N + 6$ is divisible both by 2 and 3.

In a letter to Euler, a Russian named Goldback asked if he could prove that *every even number can be written as the sum of two primes.* Mathematicians of the 18th Century communicated their discoveries one to the other by letter, as there were few journals, so that their collected works consist in large part of correspondence. As a result of this fact, this unproved proposition is known as the Goldbach conjecture, though he is known for nothing else. However, the proposition is reasonable, for

---

[1]The Euler product $\prod_p^\infty [1/(1 - 1/p)]$ is definitely divergent to infinity so that the reciprocal product $\prod_p^\infty (1 - 1/p)$ diverges to zero. $\sum_p^\infty 1/p^2$ is convergent as it is a subseries of the absolutely convergent $\sum_1^\infty 1/n^2$. Hence from the theorem, if $\sum a_n^2$ converges, then $\sum a_n$ and $\prod (1 + a_n)$ converge and diverge together, it follows that $\sum_p^\infty 1/p$ is divergent.

$$4 = 2 + 2, \qquad\qquad 12 = 5 + 7,$$
$$6 = 3 + 3, \qquad\qquad 14 = 7 + 7 = 3 + 11,$$
$$8 = 5 + 3, \qquad\qquad 16 = 5 + 11 = 3 + 13,$$
$$10 = 5 + 5 = 3 + 7, \qquad 18 = 7 + 11 = 5 + 13,$$

and no one has ever found an even number contradicting the Goldbach conjecture. However, it is unproved. A major difficulty in proving this is the nature of prime numbers — they are made for multiplication, while the proposition is of an additive nature.

If the Goldbach conjecture were true, then adding 3 to every even number we should have: Every odd number can be expressed as the sum of three primes. However, this weaker proposition does not imply the Goldbach conjecture — even if it were valid, still some even numbers might not be expressible as the sum of two primes. This has not been completely proved, but Vinogradov (1937), using ideas developed by Hardy and Littlewood in the early 1920's, succeeded in proving that *from a certain number, M, onward, all odd numbers are the sum of three primes.* Unfortunately his proof is an existence proof; it does not yield a method of estimating M. Nevertheless, the importance of this result is not to be underestimated.

Here we have a statement about *all* odd numbers greater than M — one we could never verify experimentally. A function of mathematics is to prove these things which are beyond experimental verification, and in large part, the importance and interest of mathematics lie in its "infinite tail," in those propositions which are not experimentally verifiable. For example, $5 \cdot 3 = 3 \cdot 5$ is mathematically dull, for we can check it; while $a \cdot b = b \cdot a$ is fascinating, for it is a statement about *all* numbers.

We have noticed that the primes seem to be distributed irregularly. However, Bertrand observed that *between a and 2a there always is a prime number.* For example: $1 < 2 \leq 2, 2 < 3 < 4, 3 < 5 < 6, 4 < 5, 7 < 8, 5 < 7 < 10$, etc. This is known as Bertrand's Postulate and has been proved by rather elementary means. Comparing the size of the interval, $a < p \leq 2a$, with the first number, a, we have the relative length of the interval $(2a - a)/a = 1$. Thus the primes are regularly distributed in the sense that there is at least one in each interval of relative length 1.

Now it would seem reasonable to ask if a smaller interval can be assigned in which we can always find at least one prime. For example, is the inequality $a^2 < p < (a + 1)^2$ valid for all a? This yields $1 < 2, 3 < 4, 4 < 5, 7 < 9$, for a equal 1 and 2. Here the relative length of the interval is

$$\frac{(a + 1)^2 - a^2}{a^2} = \frac{2a + 1}{a^2} = \frac{2}{a} + \frac{1}{a^2} \sim \frac{2}{a},$$

so that for large numbers the inequality, if valid, would assign a relative interval in which we could find a prime much smaller than that of Bertrand's Postulate. Unfortunately this is unproved.

However Tchudakov and Titchmarsh [see H. Montgomery (1971) for a discussion of this] have shown that *between two consecutive fifth powers, a^5 and $(a + 1)^5$, there is always a prime, provided that a is sufficiently large*. In this case the relative interval is

$$\frac{(a + 1)^5 - a^5}{a^5} = \frac{5}{a} + \frac{10}{a^2} + \frac{10}{a^3} + \frac{5}{a^4} + \frac{1}{a^5} \sim \frac{5}{a}$$

for large a, which is much smaller than the relative interval of Bertrand's Postulate. Thus the primes are distributed in such a way that we can find one in each interval of relative length, i.e., between a and $2a$, and, for sufficiently large a, of relative length of the order of $5/a$, i.e., between a^5 and $(a + 1)^5$.

Note to Chapter 1

Brun's famous theorem that the sum of the reciprocals of the twin primes converges [see E. Landau (1958)] has led to a new branch of number theory called sieve theory. The Selberg sieve [cf., C. Hooley (1976), Halberstam, Richert (1974)] and the "large sieve" [cf., E. Bombieri (1974), H. Montgomery (1978)] have been responsible for many new advances in the theory of prime numbers. Perhaps the most striking result is due to the Chinese mathematician J. Chen (1973) who proved that every sufficiently large even number can be expressed as the sum of a prime and a number having at most two prime factors. Using a new approach based on the "large sieve", H. Montgomery (1969) showed that for x sufficiently large and $\alpha = \frac{3}{5}$, there is always a prime between x and $x + x^\alpha$. This was improved to $\alpha = \frac{7}{12}$ by Huxley (1972) and more recently to $\alpha = \frac{13}{23}$ by Iwaniec and Jutila (1979). Under the assumption of the Riemann hypotheses, it is not hard to show that for sufficiently large x, there is a prime between x and $x + x^\alpha$ for any $\alpha > \frac{1}{2}$.

Chapter 2

Decomposition of Numbers into Prime Factors

In this chapter we shall consider prime numbers in their proper use — in building up numbers multiplicatively. The questions that we consider here are elementary in that they arise in high school; but there, and rightly so, the results are tacitly assumed, as they are too difficult. But a teacher should understand these things, if only to avoid problems that may arise.

We can decompose the number 60 in the following ways: $60 = 4 \cdot 15 = 2 \cdot 2 \cdot 3 \cdot 5$, or $60 = 3 \cdot 20 = 3 \cdot 4 \cdot 5 = 3 \cdot 2 \cdot 2 \cdot 5$, or $60 = 6 \cdot 10 = 2 \cdot 3 \cdot 2 \cdot 5$, etc., in each of which we represent 60 as a rearrangement of the product of the prime numbers, $2 \cdot 2 \cdot 3 \cdot 5$. (The special feature of multiplication, namely, that the order of the factors makes no difference, is usually called the commutative law of multiplication.) To us it is intuitively evident that the decomposition into prime factors is unique — and, indeed, that it is always possible.

Logically, a prime number is one that is no longer decomposable into other numbers.

Let us consider the realm of all numbers of the form $3n + 1$, i.e., the numbers:

$$1, 4, 7, 10, 13, 16, 19, 22, 25, 28, \ldots, 3n + 1, \ldots$$

The product of any two of these numbers,

$$(3l + 1) \cdot (3m + 1) = 9lm + 3m + 3l + 1$$
$$= 3(3lm + l + m) + 1$$
$$= 3K + 1,$$

is again a number of the same sort. Thus we can multiply in this realm without mishap, without leaving it. Here 28 is a composite number as $28 = 4 \cdot 7$ and both 4 and 7 belong to our list. Similarly $16 = 4 \cdot 4$ is also a composite number. On the other hand 10 is a number of the sort $3n + 1$ which is no longer decomposable into other numbers of the same kind, i.e., it is here a prime.

The number 100 belongs to this realm and we can decompose it as $100 = 10 \cdot 10$, or $100 = 4 \cdot 25$, where 4, 10, and 25 are numbers no longer decomposable into other numbers of the form $3n + 1$. Thus 100 may be decomposed into factors which are prime in this realm in two different ways.

If the notion of indecomposability would entail logically that the decomposition were unique, then we would find that the decomposition of a number of the form $3n + 1$ would certainly be so. Hence this counter-example shows that it is indeed necessary to prove that for ordinary numbers the decomposition into prime factors is unique.

But first let us consider another more famous and more involved counter-example. Here we will speak of complex numbers of the form $a + b\sqrt{-5}$, and we will call such a number an "integer" if a and b are ordinary integers, e.g., in this realm of numbers $3 + 2\sqrt{-5}$ is an integer. Applying the operations of addition and multiplication to these numbers, we see that the result is also a number of the same sort:

<table>
<tr><td align="center">Addition</td><td align="center">Multiplication</td></tr>
<tr><td align="center">$a + b\sqrt{-5}$</td><td align="center">$a + b\sqrt{-5}$</td></tr>
<tr><td align="center">$c + d\sqrt{-5}$</td><td align="center">$c + d\sqrt{-5}$</td></tr>
<tr><td align="center">$(a + c) + (b + d)\sqrt{-5}$</td><td align="center">$ac + bc\sqrt{-5}$</td></tr>
<tr><td></td><td align="center">$+ ad\sqrt{-5} + bd(-5)$</td></tr>
<tr><td></td><td align="center">$ac + (ad + bc)\sqrt{-5} - 5bd$</td></tr>
<tr><td></td><td align="center">$= (ac - 5bd) + (ad + bc)\sqrt{-5}$</td></tr>
</table>

And in particular we see that integers go into integers by these operations. Thus this example is better than the previous one in that here the realm is preserved both in addition as well as in multiplication.

Here 21 can be decomposed in the following ways:

$$21 = 3 \cdot 7$$
$$= \left(1 + 2\sqrt{-5}\right) \cdot \left(1 - 2\sqrt{-5}\right)$$
$$= \left(4 + \sqrt{-5}\right) \cdot \left(4 - \sqrt{-5}\right).$$

Now are these prime factors, or are they subject to further decomposition? Suppose

$$3 = \left(a + b\sqrt{-5}\right)\left(c + d\sqrt{-5}\right)$$
$$= (ac - 5bd) + (ad + bc)\sqrt{-5}.$$

If 3 were decomposable, then $(ad + bc) \equiv 0$. But if this is true, then also we have

$$3 = \left(a - b\sqrt{-5}\right)\left(c - d\sqrt{-5}\right)$$
$$= (ac - 5bd) - (ad + bc)\sqrt{-5},$$

so that

$$9 = \left(a + b\sqrt{-5}\right)\left(a - b\sqrt{-5}\right)\left(c + d\sqrt{-5}\right)\left(c - d\sqrt{-5}\right)$$
$$= (a^2 + 5b^2)(c^2 + 5d^2),$$

in which a, b, c, and d are ordinary integers. Clearly this relation is satisfied by $a = 2$, $b = 1$, $c = 1$, and $d = 0$, or by $a = 1$, $b = 0$, $c = 2$, and $d = 1$; but neither of these solutions makes sense when introduced into the original equations, $3 = (a + b\sqrt{-5})(c + d\sqrt{-5})$ or $3 = (a - b\sqrt{-5})(c - d\sqrt{-5})$. Hence 3 cannot be decomposed into further factors in this realm.

We can make such a proof more systematic by introducing the notion of the norm of a number. A product such as $(4 + \sqrt{-5})(4 - \sqrt{-5})$ has a value free of imaginary terms, or more generally

$$(a + b\sqrt{-5})(a - b\sqrt{-5}) = a^2 - (b\sqrt{-5})^2 = a^2 + 5b^2.$$

This product of $(a + b\sqrt{-5})$ by its conjugate, $(a - b\sqrt{-5})$, we call the norm of $a + b\sqrt{-5}$, which we can write

$$N(a + b\sqrt{-5}) = a^2 + 5b^2.$$

Obviously $N(a + b\sqrt{-5}) = N(a - b\sqrt{-5})$. And we have

$$
\begin{aligned}
N\{(a + b\sqrt{-5})(c + d\sqrt{-5})\} &= N\{(ac - 5bd) + (ad + bc)\sqrt{-5}\} \\
&= \{(ac - 5bd) + (ad + bc)\sqrt{-5}\} \cdot \{(ac - 5bd) - (ad + bc)\sqrt{-5}\} \\
&= (a + b\sqrt{-5})(c + d\sqrt{-5})(a - b\sqrt{-5})(c - d\sqrt{-5}) \\
&= (a + b\sqrt{-5})(a - b\sqrt{-5}) \cdot (c + d\sqrt{-5})(c - d\sqrt{-5}) \\
&= N(a + b\sqrt{-5})N(c + d\sqrt{-5}).
\end{aligned}
$$

so that the norm of a product is the product of the norms.

Now suppose that $(4 + \sqrt{-5}) = (x + y\sqrt{-5})(u + v\sqrt{-5})$: then we have $N(4 + \sqrt{-5}) = N(x + y\sqrt{-5})N(u + v\sqrt{-5})$, or $4^2 + 5 = 21 = (x^2 + 5y^2)$ $\cdot (u^2 + 5v^2)$, where x, y, u, and v are ordinary numbers. But $21 = 21 \cdot 1$ or $21 = 3 \cdot 7$. If $x^2 + 5y^2 = 1$, then as x and y are integers, $x = 1$ and $y = 0$. Hence $u^2 + 5v^2 = 21$, so that $u = 4$ and $v = 1$, or $u = 1$ and $v = 2$, are the only integers that satisfy this. The latter makes no sense in the present case while $u = 4$ and $v = 1$ is no decomposition at all. If $x^2 + 5y^2 = 3$, or if $x^2 + 5y^2 = 7$, then there are no solutions in integers for x and y. Thus we are forced to conclude that $4 + \sqrt{-5}$ and $4 - \sqrt{-5}$ are prime factors of 21 in this realm, since they are no longer decomposable.

Since we have already shown that 3 is not decomposable here, let us next turn to 7. Suppose $7 = (x + y\sqrt{-5})(u + v\sqrt{-5})$

$$N(7) = 49 = N(x + y\sqrt{-5})N(u + v\sqrt{-5}) = (x^2 + 5y^2)(u^2 - 5v^2).$$

Now $49 = 1 \cdot 49$ or $49 = 7 \cdot 7$. If $x^2 + 5y = 1$, $x = 1$, $y = 0$, and again we see that this leads to no decomposition, while if $x^2 + 5y^2 = 7$, there are no solutions in integers.

Thus, in the foregoing, we have shown that in this realm 21 can be decomposed into factors that are no longer decomposable, i.e., into prime factors, in at least two ways. Hence we have an example of a realm of numbers in which we can add, subtract, multiply — and decompose the numbers into prime factors that are not unique.

Hence uniqueness of decomposition is not a logical consequence of the possibility of decomposition. And we must make a proof of the fact that the decomposition of ordinary integers is unique.

Now we can ask what is the special reason for the uniqueness of the prime factorization of ordinary integers? What property do they possess that makes the proof of this possible? Curiously, all the elements of the proof are contained in Euclid. The genius of the Greeks as mathematicians seems even greater when we realize that they recognized the necessity of such a proof and made it without having counter-examples such as we have just produced.

We know what a common multiple is. 24 and 60 have many, e.g., $24 \times 60 = 1440$, etc. From two numbers a and b, we form $a \cdot b$ which is certainly a common multiple of a and b; and from this we can deduce many more. But in the case of 24 and 60, we need only a number as large as 120 to find a common multiple, as $120 = 2 \cdot 60 = 5 \cdot 24$. Thus the existence of at least one common multiple cannot be questioned, and since there are only a finite number of integers less than $a \cdot b$, it is clear that we can always find a least common multiple (LCM) given a and b. The least common multiple plays an important role among all the common multiples, as any other common multiple must be a multiple of it.

Let us write $\mu(a, b) = $ LCM of a and b. *If M is a common multiple of a and b, then $M = h \cdot \mu(a, b)$.*

Proof. If $M = \mu(a, b)$, then there is nothing to prove. Therefore we assume that $M > \mu(a, b)$. Performing division, we have

$$M = q\mu + r$$

where $0 \le r < \mu$ which we write as $r = M - q\mu$. Now we wish to show that r must be zero. However, r is in fact a common multiple of a and b, for $M = \alpha a = \gamma b$ and $\mu = \beta a = \delta b$, so that $r = a(\alpha - q\beta) = b(\gamma - q\delta)$. Hence r is a common multiple of a and b and is also less than μ, as $0 \le r < \mu$. But since μ is the least common multiple, this is impossible unless $r = 0$. Whence $M = q\mu$, which is the assertion that $M = h \cdot \mu(a, b)$.

Now comes the most important part of the proof. Assume we are given p, a prime, which by definition has only the divisors 1 and p and also a product $a \cdot b$ that is a multiple of p, i.e., $a \cdot b = \lambda \cdot p$; or $a \cdot b$ is divisible by p; then either a or b is divisible by p. This we express more concisely in the theorem: *If a product is divisible by a prime, then at least one of the factors is divisible by the prime.*

From the outset we may assume a is not divisible by p, so that we must show b to be a multiple of p.

Let $\mu(a, p) = \mu$ be the least common multiple of a and p. As $a \cdot b$ is also a common multiple of a and p, we have $a \cdot b = h \cdot \mu$.

Now $a \cdot p = k \cdot \mu$, or $a = k \cdot \mu/p = k \cdot \rho$ and $p = k \cdot \mu/a = k \cdot \sigma$, ρ and σ being integers, so that both a and p are multiples of k. However, p is prime so that $k = 1$ or p. But $k \ne p$; for if $k = p$, then $a = p \cdot \rho$, or a would be divisible by p, which is contrary to our assumption. Hence $k = 1$, or $a \cdot p = \mu$. Thus $a \cdot b = h \cdot a \cdot p$, so that $b = h \cdot p$, and b is indeed divisible by p.

This theorem clearly does not hold in each of our counter-examples, where if the product $a \cdot b$ is a multiple of a prime p, it does not follow that either a or b is divisible by p. For example, in the realm $a + b\sqrt{-5}$ where $21 = 3 \cdot 7 = (4 + \sqrt{-5})(4 - \sqrt{-5})$, 21 is a multiple of each of the factors, but no one of them is divisible by another. This may also be verified for the realm of integers of the form $3n + 1$.

As indicated in the enunciation, the theorem can be extended to larger numbers of factors, e.g., *if $a \cdot b \cdot c$ is divisible by p, then at least one of a, b, c must be divisible by p.* This follows from the iteration of the theorem for two factors. From the outset let us assume that p does not divide a, but does divide $abc = a(bc)$. Then by the preceding, p must divide bc. And applying the theorem again, we see that p must divide either b or c. Hence if p divides abc, it must divide a, or b, or c. And by such iteration we can extend the theorem to as many factors as we choose.

Finally, we prove the essential theorem of this chapter, namely *the decomposition of ordinary integers into prime factors is unique.*

Suppose that we have decomposed the number n into prime factors in two ways, so that $n = p_1 p_2 \cdots p_l = q_1 q_2 \cdots q_k \cdot n$ is divisible by p_1, hence $q_1 q_2 \cdots q_k$ is divisible by p_1. However, by the foregoing at least one factor is divisible by p_1, say q_1. Since the q's are primes and the p's are primes, $q_1 = p_1$.

Thus $p_2 p_2 \cdots p_l = q_1 q_2 \cdots q_k$. And repeating the foregoing process, we find that the q's and p's are indeed identical, so that the prime number factorization is unique in the realm of ordinary integers.

Note to Chapter 2

The problem of unique factorization was a central issue among mathematicians of the last century. In 1847, Lamé announced at the Paris academy that he had found a proof of Fermat's famous last theorem which says that $x^n + y^n = z^n$ is impossible in positive integers x, y, z if $n > 2$ is an integer. His idea was based on the following factorization. Let $\theta = \cos(2\pi/n) + i \sin(2\pi/n) = e^{2\pi i/n}$. Then

$$x^n + y^n = (x + y)(x + \theta y) \cdots (x + \theta^{n-1} y) \qquad (n \text{ odd}).$$

Now, Lamé thought that if the factors $x + y, x + \theta y, \ldots, x + \theta^{n-1} y$ are pairwise relatively prime in the domain of all cyclotomic integers

$$a_0 + a_1 \theta + \cdots + a_{n-1} \theta^{n-1} \qquad (a_i \text{ integers}),$$

then each of the factors $x + y, x + y, x + \theta y, \ldots$ must itself be an nth power. In this manner he hoped to obtain a contradiction. As Rademacher beautifully shows in this chapter, the law of unique factorization may not be valid in arbitrary domains of integers of various types. In fact, it can be shown that unique factorization fails where Lamé thought it was true. A correct proof of Fermat's last theorem is unknown to this day.

Chapter 3

Common Fractions

In order to introduce common fractions into our number system, we take for granted that we know all the integers, positive and negative, the operations, addition and multiplication, that we can perform on them, and the rules by which we may combine these operations. As we become more sophisticated in mathematics, we find a need to enlarge our number system — from the integers to the rational numbers, to the irrational numbers, and to the complex numbers. But this is a heuristic approach; whereas a more formal approach is desirable in order to show that the operations performed with fractions fit the same pattern as the operations with integers.

What are these rules of operation that we take for granted? They seem to have been first enumerated by the Irish mathematician William Hamilton.

We have a *commutative law*:
for addition: $a + b = b + a$,
for multiplication: $ab = ba$;

an *associative law*:
for addition: $a + (b + c) = (a + b) + c$,
for multiplication: $a(bc) = (ab)c$;

(the associative law for addition is merely codification of the fact that we can only add two numbers together, e.g., the method we use in adding long columns of numbers);

and a *distributive law*: $a \cdot (b + c) = ab + ac$,

which connects addition and multiplication. This last law is clearly of a different nature from the first two in which we pass from the law for addition to the law for multiplication by replacing the plus sign by a dot. If we apply this replacement to the distributive law, we have nonsense: $a + (b \cdot c) \neq (a + b) \cdot (a + c)$. The usual rule of multiplication is a clear application of the distributive law, e.g.,

$$14 \cdot 26 = 14(20 + 6) = 14 \cdot 20 + 14 \cdot 6$$
$$= 280 + 84 = 364,$$

or

$$
\begin{array}{r}
14 \\
26 \\
\hline
84 \\
28 \\
\hline
364\,.
\end{array}
$$

Finally we have a *law of cancellation*:

for addition: if $a + b = a + c$, then $b = c$.
for multiplication: if $ab = ac$, and if $a \neq 0$, then $b = c$.

In addition to these laws we have two *invariants*, one for each operation:

for addition: zero, i.e., $a + 0 = a$.
for multiplication: one, i.e., $a \cdot 1 = a$.

The foregoing are postulates and from them we can derive the other familiar rules of arithmetic. For example we have as a theorem: $a \cdot 0 = 0$.

Proof

$$a \cdot b = a \cdot (b + 0) \qquad \text{(the invariant for addition)},$$
but $a \cdot (b + 0) = a \cdot b + a \cdot 0$ or
$$a \cdot b = a \cdot b + a \cdot 0 \qquad \text{(the distributive law)}$$
whence $a \cdot 0 = 0$ (the law of cancellation).

Knowing the integers and having these rules of operation in mind, we may now introduce fractions as an ordered pair of integers $(a|b)$, where $b \neq 0$. In this ordered pair a plays the conventional role of the numerator, while b plays the role of the denominator. We introduce purposely a new symbol $(a|b)$ in order to be sure that we do not inadvertntly make use of the rules of fractions which we are now going to introduce.

Definitions

Multiplication of Fractions: $(a|b)(c|d) = (ac|bd)$,
Addition of Fractions: $(a|b) + (c|d) = (ad + bc|bd)$.

In addition to these two operations we need an equivalence relation — when are two pairs of numbers equal? — for among all the ordinary fractions there are many that are equal, e.g., $\frac{3}{4} = \frac{6}{8} = \frac{15}{20} = \frac{-9}{-12} = \frac{-12}{-16}$, etc.

Definition. Equivalence of Fractions: $(a|b) = (c|d)$, if and only if $ad = bc$.

This definition of equivalence is based on the known operations of the integer realm and reduces the decision of equivalence to a question therein.

An equivalence relation must have three properties:

1. *Reflexivity*: every pair must be equal to itself. $(a|b) = (a|b)$ implies $ab = ba$ which is true in the integer realm.
2. *Symmetry*: if a pair is equal to a second, the second pair is equal to the first. $(a|b) = (c|d)$ implies $ad = bc$ which can be written as $cb = da$. But the latter is merely the integer formulation of $(c|d) = (a|b)$.
3. *Transitivity*: if a pair is equal to a second, and the second is equal to a third, then the first pair is equal to the third. If $(a|b) = (c|d)$ and $(c|d) = (e|f)$, then $(a|b) = (e|f)$. These imply

$$ad = bc, \qquad cf = de,$$
$$adf = bcf, \qquad bcf = bde,$$

whence

$$adf = bde,$$

and by cancellation (since $d \neq 0$)

$$af = be,$$

so that $(a|b) = (e|f)$. And we have shown that our proposed equivalence relation indeed does have the properties required of it.

In so defining equivalence between fractions, we have divided them into classes such that $\frac{3}{4}$ and $\frac{12}{16}$ belong to the same class. Here the essential difficulty with fractions is exhibited, namely that we have no unique notation for the classes.

Suppose that we have two fractions of the same class, i.e., $(a|b) = (a'|b')$; then it would seem reasonable to assert that $(a|b)(c|d) = (a'|b')(c|d)$, or *multiplication by a member of a class is equivalent to multiplication by any other member of the same class*. $(a|b) = (a'|b')$ implies that $ab' = a'b$; and we must show that $(a|b)(c|d) = (a'|b')(c|d)$, or $(ac|bd) = (a'c|b'd)$, which implies that $acb'd = a'cbd$. But this last relation holds in the realm of integers.

Likewise, we could show that *the addition of a member of a class is equivalent to the addition of any other member of the same class*. And similarly, the commutative, associative, and distributive laws also hold for fractions as we have defined them. These are left as exercises.

Like 0 among the integers *there is a class of fractions invariant with respect to addition, namely* $(0|a) = (0|b) = (0|1)$, *provided that a and b are not zero*. To show this we note that $(a|b) + (0|d) = (ad + b \cdot 0|bd) = (ad|bd) = (a|b)$, for the last equality implies $adb = bda$. Similarly we find *a class of fractions invariant with respect to multiplication, namely* $(a|a) = (b|b) = (1|1)$, *provided a and b are not zero*, for $(a|a) \cdot (b|c) = (ab|ac) = (b|c)$.

Now we can show that *there is a fraction* $(x|y)$ *such that* $(a|b) \cdot (x|y) = (c|d)$, *provided that neither a, b or d are zero*. $(a|b) \cdot (x|y) = (ax|by) = (c|d)$ can be fulfilled by $ax = kc$ and $by = kd$. If we put $k = ab$, so that $ax = abc$ and

$by = abd$, then by the law of cancellation we have $x = bc$ and $y = ad$. Thus the fractions always permit division, provided that of all the terms only the first term of the dividend may be zero.

Among these fractions there is a subset of classes of the form $(a|1)$, such that $(a|1) \cdot (b|1) = (ab|1)$ and $(a|1) + (b|1) = (a \cdot 1 + 1 \cdot b|1 \cdot 1) = (a + b|1)$. This subset of number pairs is isomorphic with the set of integers, i.e., $(a|1) \leftrightarrow a$, $(b|1) \leftrightarrow b$, $(a|1) \cdot (b|1) \leftrightarrow ab$. When we carry out an operation on the fractions and on the corresponding integers under this isomorphic relation, we find, after completing the operation, that the correspondence is preserved. Thus *the wider realm of fractions contains the integers as a subset and moreover always permits division.*

In mathematics, we frequently classify things and then single out from each class an element, called the *reduced element*, to stand for the whole class. This process proves so useful that when it is impossible to exhibit a method for finding a reduced element, to avoid the difficulty we postulate the existence of such an element as in the axiom of choice in the theory of sets.

Our classes consist of equivalent fractions $[(a|b) = (c|d)$, if and only if $ad = bc]$ and we choose as the reduced element that pair of the class which has the smallest positive second element (denominator). Of all the fractions of a class, only one can be the reduced element. (Suppose there were two, then $(a|b) = (c|b)$, as they must have the same second element, which implies $ab = cb$, so that $a = c$.) We do not need to make an involved proof of the existence of the reduced pair. Suppose we have our class of fractions in a bag. Pulling one from the bag we note its denominator. If it is negative, changing the signs of both members of the pair will give us a fraction with positive denominator which also belongs to our class. Now we have only to make a finite number of comparisons to determine if a smaller positive number can serve as a denominator of a fraction of our class. Since there can be only a finite number of smaller denominators, we can certainly determine the smallest by inspection. When only a finite number of objects are to be compared, inspection is indeed a legitimate and feasible method of mathematical proof. Thus *we can always find one fraction, $(a|b)$, the reduced element of the class, such that b is the smallest positive denominator of the equivalent fractions.*

In what sense does this element represent the class? We shall show that *if $(a|b)$ is the reduced fraction of a class and $(c|d)$ belongs to the same class, then $(c|d) = (ka|kb)$.*

Consider $(a|b) = (c|d)$, where $b > 0$, $d > 0$, and $d \geq b$, where b is the smallest positive denominator of all equivalent fractions. $(c|d)$ is a fraction different from $(a|b)$, only if $d > b$; for if $d = b$, then $a = c$, as $ad = bc = dc$. So let us suppose definitely that $d > b$. Performing division we have $d = q \cdot b + r$, where $0 < r \leq b$. (Note the curious specification of the remainder r — zero is excluded, while b is included as a possible value.) Set $c = q \cdot a + s$, where q is defined by the previous division. We know that $(a|b) = (c|d) = (qa + s|qb + r)$, implying that $a(qb + r) = b(qa + s)$, or $ar = bs$: whence $(a|b) = (s|r)$. But as b was the smallest positive denominator contained in the class, it is impossible that r be different from b, i.e., $r = b$. This implies that $d = qb + b = (q + 1)b$. Now we must show that $s = a$; but this follows from $(a|b) = (s|r) = (s|b)$, so that $c = (q + 1)a$. And thus $(c|d) = [(q + 1)a|(q + 1)b] = (ka|kb)$.

We have defined the reduced fraction as the member of the class that has the smallest positive denominator, and we have shown that any other member of the class is merely an amplification of the reduced fraction in which the numerator and denominator are both multiplied by the same factor. This leads naturally to another definition of the reduced fraction, one that is more convenient: $(a|b)$ *is reduced if and only if a and b have no common divisor other than* ± 1. We must show this to be equivalent to the first definition. First suppose that a and b have a common divisor, i.e., $a = \delta\alpha$, $b = \delta\beta$, where $\delta \neq \pm 1$. Then $(a|b) = (\delta\alpha|\delta\beta) = (\alpha|\beta)$, $b > \beta$, so that $(a|b)$ is not a reduced fraction in the old sense. Conversely, if c and d have no common divisor, the $(c|d)$, $d > 0$, must be a reduced fraction. Indeed, if it were not reduced, then there would be another specimen in the same class with a smaller positive denominator, say $(a|b)$. But we just showed that $(c|d) = (ka|kb)$ and by hypothesis $k = \pm 1$, so that $(c|d)$ is in fact the reduced fraction.

The notion that a and b have no common divisor is perhaps more familiarly expressed by saying that the fraction is in lowest terms. For example, consider

$$\frac{31\,759}{112\,753} = \frac{\cdots\cdots}{\cdots\cdots}.$$

which may or may not be in lowest terms. Why should it be impossible that two fractions, both in lowest terms, should be equal? Thinking over the example, what it implies in the realm of integers, we see that this impossibility is equivalent to the theorem on the uniqueness of prime factorization of a number. Thus our next goal will be to establish this in full. To this end we shall use the previous result to prove Euclid's lemma.

Suppose a and b are coprime, i.e., $(a, b) = 1$, *and suppose a divides the product bc, then a divides c.* Let us write out what a divides bc means. That is $bc = ad$, or $ad = bc$, which can be written in fractional notation as $(a|b) = (c|d)$, where $(a|b)$ is a reduced fraction. Hence $c = ak$, or c is divisible by a.

By a specialization of Euclid's lemma we have: *If p is a prime, and p divides bc, then p divides either b or c.*

1) Suppose p divides b, then the theorem is true.
2) Suppose p does not divide b, then $(p, b) = 1$, for 1 is a divisor of both and p has only two divisors, 1 and p, so that (p, b), the greatest common divisor, is in fact equal to 1. Hence by Euclid's lemma p divides c.

But this is tantamount to the theorem on the uniqueness of prime factorization of integers, being the essential theorem used in the proof of that important proposition.

The notation $(a|b)$ for common fractions and the abstractness of the presentation in this chapter are certainly not to be recommended for the teaching of fractions.

Rather they were used to remove, in so far as possible, all previous associations that we have with fractions and their operations in order to exhibit clearly that they possess intrinsically the same properties as the integers.

Chapter 4

Order of Fractions

So far we have used only the notion of equality between fractions and integers. However, the integers are ordered, for if we are given any two integers a and b, we find one of three relations between them: Either

$$a < b, \quad a = b, \quad \text{or} \quad a > b.$$

Since $b > a$ implies $a < b$, we consider in detail only the relation less than.

Inequality, $a < b$, is a non-reflexive, asymmetric, but transitive relation. In the previous chapter we showed that equality is transitive. Another example is the similarity of two triangles — which is also reflexive, as $A \sim A$, and symmetric, for if $A \sim B$, then $B \sim A$ — a relation exhibiting the property of transitivity: If $A \sim B$ and $B \sim C$, then $A \sim B$. Inequality is likewise transitives: If $a < b$ and $b < c$, then $a < c$. This is an ordering which holds for the whole scale of integers. The foregoing we take for granted.

Next we show that fractions can also be ordered. Let us revert to the more familiar notation a/b. We can always assume that the denominator of a fraction a/b is positive, for multiplying numerator and denominator by -1, we obtain an equivalent fraction. With this convention we introduce the notion of the inequality of fractions:

Definition. Inequality of Fractions: $a/b < c/d$, where b, $d > 0$, means $ad < cb$.

The relation of inequality between two fractions is clearly non-reflexive, $a/b < a/b$ is nonsense, and asymmetric. Thus we have a trichotomy of relations: Given any two fractions a/b and c/d, where b, $d > 0$, either $a/b < c/d$, $a/b = c/d$, or $a/b > c/d$, as $ad < bc$, or $ad = bc$, or $ad > bc$. Hence fractions are ordered.

An inequality between integers remains true if we multiply both sides by a positive number, e.g.:

$$-3 < 5 \quad \text{or} \quad x < y, \quad k > 0,$$
$$-3.4 < 5.4, \quad kx < ky.$$

But this does not hold for negative numbers. The rule has a complement for addition: If $x < y$, the $x + z < y + z$, where z may be any integer.

With the help of these rules, we can prove the transitivity of the inequality of fractions directly from the definition.

Suppose $a/b < c/d$ and $c/d < e/f$; where b, c, $f > 0$, which means $ad < bc$, $cf < de$, and as b, $f > 0$, we have $adf < bcf$, $bcf < bde$, or $adf < bde$, and since $d > 0$, we can cancel it, i.e., $af < be$, which means that $a/b < e/f$. Thus the transitivity of the inequality between fractions is proved.

Fractions obey the rules for inequalities between integers. For example, we prove from our definitions that : If $m/n < q/r$, n, $r > 0$, then $m/n + u/v < q/r + u/v$, where $v > 0$. All we have to do is to reduce these things to their meaning. The inequality which we want to establish means:

$$\frac{mv + nu}{nv} < \frac{qv + ru}{rv}$$

which should be a consequence of the things we have already mentioned. Thus $(mv + nu)rv < nv(qv + ru)$, and since $v > 0$, we can cancel, yielding $(mv + nu)r < n(qv + ru)$, or

$$mrv + nru < nqv + nru,$$

and by the rule of addition for inequalities we have

$$mrv < nqv,$$

but as $v > 0$, we have on cancelling $mr < nq$, which is our original assumption.

Similarly we should be able to prove that: If $m/n < q/r$, n, $r > 0$, and if $u/v > 0$, $v > 0$, then $m/n \cdot u/v < q/r \cdot u/v$. But what does $a/b > 0$, $b > 0$ mean? This we can write as $a/b > 0/1$, which implies $a > 0$. The completion of this proof is left as an exercise. These elementary things we now take for granted and proceed to the main material of this chapter, Farey sequences.

Farey, an English geologist, made some remarkable observations on the properties of ordered fractions which were published in the *Philosophical Magazine* in 1816. Cauchy saw Farey's statement, supplied the proof, and named these sequences.

Let us consider only reduced fractions, say for definiteness in the interval $0 \le a/b \le 1$, although this restriction is not necessary. Let us write all fractions with denominator ≤ 1 in their proper order:

$$\frac{0}{1} < \frac{1}{1}$$

Now admit denominators 2,

$$\frac{0}{1} < \frac{1}{2} < \frac{1}{1}$$

and denominators 3

$$\frac{0}{1} < \frac{1}{3} < \frac{1}{2} < \frac{2}{3} < \frac{1}{1}$$

and denominators 4

$$\frac{0}{1} < \frac{1}{4} < \frac{1}{3} < \frac{1}{2} < \frac{2}{3} < \frac{3}{4} < \frac{1}{1}.$$

Such a sequence is called a *Farey sequence*. And the arrangement of all the reduced fractions $0 \leq a/b \leq 1$, with denominator $\leq N$ in their proper order is called the *Farey sequence of order N*.

Farey made the following observation. We can speak of adjacent fractions in such a sequence, i.e., when $N = 4$, $\frac{1}{2}$ and $\frac{2}{3}$ are adjacent, as are $\frac{2}{3}$ and $\frac{3}{4}$. Forming the difference of these

$$\frac{2}{3} - \frac{1}{2} = \frac{1}{2 \cdot 3}$$

$$\frac{3}{4} - \frac{2}{3} = \frac{1}{3 \cdot 4},$$

we see that this difference is of the form one divided by the product of the denominators. Farey claimed that all such differences were of this form.

Let $a/b < c/d$ be adjacent in a Farey sequence. From these two fractions we can form what Hardy and Littlewood call the "mediant" of these fractions, i.e., $(a + c)/(b + d)$, the sum of the numerators divided by the sum of the denominators. (Don't teach this in school, for some student is sure to confuse it with the addition of fractions.) The mediant has the following property:

$$\frac{a}{b} < \frac{a + c}{b + d} < \frac{c}{d}, \qquad b, d > 0.$$

We verify the left hand inequality. $a/b < (a + c)/(b + d)$ means $a(b + d) < b(a + c)$, or $ab + ad < ab + bc$, or $ad < bc$, which is equivalent to our assumption that $a/b < c/d$. And we can also verify the right hand inequality. Thus *the mediant lies between the two original fractions*.

By forming the mediant of adjacent fractions of the Farey sequence of order $N - 1$ and inserting them in their proper place, we obtain the Farey sequence of order N, provided we neglect the mediants whose denominators are greater than N. For example, for $N - 1 = 3$, the mediant of $\frac{1}{3}$ and $\frac{1}{2}$, namely $\frac{2}{5}$, belongs to the Farey sequence of order 5, rather than to that of order 4 (cf., Plate II).

Thus we see that this method of constructing Farey sequences yields fractions of the proper order. However we still must show that this process produces the other Farey properties, namely, all proper fractions appear in reduced form and the difference between two adjacents is 1 divided by the product of the denominators. Specifically it must be shown that:

$$(N = 1) \quad \tfrac{0}{1} < \tfrac{1}{1}$$

$$(N = 2) \quad \tfrac{0}{1} < \tfrac{1}{2} < \tfrac{1}{1}$$

$$(N = 3) \quad \tfrac{0}{1} < \tfrac{1}{3} < \tfrac{1}{2} < \tfrac{2}{3} < \tfrac{1}{1}$$

$$(N = 4) \quad \tfrac{0}{1} < \tfrac{1}{4} < \tfrac{1}{3} < * < \tfrac{1}{2} < * < \tfrac{2}{3} < \tfrac{3}{4} < \tfrac{1}{1}$$

$$(N = 5) \quad \tfrac{0}{1} < \tfrac{1}{5} < \tfrac{1}{4} < * < \tfrac{1}{3} < \tfrac{2}{5} < \tfrac{1}{2} < \tfrac{3}{5} < \tfrac{2}{3} < * < \tfrac{3}{4} < \tfrac{4}{5} < \tfrac{1}{1}$$

$$(N = 6) \quad \tfrac{0}{1} < \tfrac{1}{6} < \tfrac{1}{5} < * < \tfrac{1}{4} < * < \tfrac{1}{3} < * < \tfrac{2}{5} < * < \tfrac{1}{2} < * < \tfrac{3}{5} < * < \tfrac{2}{3} < * < \tfrac{3}{4} < * < \tfrac{4}{5} < \tfrac{5}{6} < \tfrac{1}{1}$$

$$(N = 7) \quad \tfrac{0}{1} < \tfrac{1}{7} < \tfrac{1}{6} < * < \tfrac{1}{5} < * < \tfrac{1}{4} < \tfrac{2}{7} < \tfrac{1}{3} < * < \tfrac{2}{5} < \tfrac{3}{7} < \tfrac{1}{2} < \tfrac{4}{7} < \tfrac{3}{5} < * < \tfrac{2}{3} < \tfrac{5}{7} < \tfrac{3}{4} < * < \tfrac{4}{5} < * < \tfrac{5}{6} < \tfrac{6}{7} < \tfrac{1}{1}$$

(* indicates that a mediant belonging to a sequence of higher order has been neglected.)

Plate II

1. all proper fractions can be obtained by constructing the Nth Farey sequence from the N-1th through the formation of mediants;
2. all these fractions are reduced; and
3. the difference between the adjacent fractions is of the asserted form.

Let us assume that all of these assertions have been verified up to the N-1th sequence. We begin by proving the last assertion.

Suppose that the difference between two adjacent fractions in this sequence, $h_1/k_1 < h_2/k_2$, is of the form $h_2/k_2 - h_1/k_2 = 1/k_1k_2$. This means that $(h_2k_1 - h_1k_2)/k_1k_2 = 1/k_1k_2$, or $h_2k_1 - h_1k_2 = 1$. Now we want to show that the mediant will also share this property. From $h_1/k_1 < (h_1 + h_2)/(k_1 + k_2) < h_2/k_2$, the assertion can be written as $k_1(h_1 + h_2) - h_1(k_1 + k_2) = 1$, with a similar relation for h_2 and k_2. We write the assumption $h_2k_1 - h_1k_2 = 1$ in the form of a determinant

$$\begin{vmatrix} h_1 & h_2 \\ k_1 & k_2 \end{vmatrix} = -1.$$

Similarly we write the assertion as

$$\begin{vmatrix} h_1 & h_1 + h_2 \\ k_1 & k_1 + k_2 \end{vmatrix} = -1.$$

But the last determinant is merely a consequence of the first, as the second column is the sum of the columns of the first determinant. And the extension to $h_2/k_2 - (h_1 + h_2)/(k_1 + k_2)$ is obvious. Thus if two adjacent fractions, $h_1/k_1 < h_2/k_2$, have the property that their difference is of the form $1/k_1k_2$, then their differences with their mediant, $(h_1 + h_2)/(k_1 + k_2)$, are $1/k_1(k_1 + k_2)$ and $1/k_2(k_1 + k_2)$, or briefly, *the mediant preserves the form of the difference between adjacent fractions*.

From the determinants of the preceding proof we can easily show that *if two adjacent fractions are reduced, then their mediant must also be reduced*. For if $(h_1 + h_2)/(k_1 + k_2)$ were not reduced, we could remove a factor from the second column of the determinant

$$\begin{vmatrix} h_1 & h_1 + h_2 \\ k_1 & k_1 + k_2 \end{vmatrix} = -1.$$

But this would imply that

$$\begin{vmatrix} h_1 & h_2 \\ k_1 & k_2 \end{vmatrix},$$

which is equal to the previous determinant, could not be equal to -1, contrary to the assumption that h_1/k_1 and h_2/k_2 were adjacent.

Finally can we obtain all proper fractions by this process? Suppose a/b is a reduced fraction such that $0/1 < a/b < 1/1$ (the Farey properties can be shown for larger intervals), which means $0 < a < b$. Does this fraction, a/b, appear in the Farey sequence of order b when we construct it by forming the mediants of the sequence of order $b - 1$? The latter cannot contain a/b as all the denominators are less than b. Since all fractions are ordered, a/b will have a place between two members of the b-1th sequence, say $h_1/k_1 < a/b < h_2/k_2$, where $h_1/k_1 < h_2/k_2$ are adjacent fractions, i.e.,

$$\begin{vmatrix} h_1 & h_2 \\ k_1 & k_2 \end{vmatrix} = -1.$$

Forming the difference $a/b - h_1/k_1 = (ak_1 - bh_1)/bk_1 = d/bk_1$, we have $k_1a - h_1b = d_1$, where clearly $d_1 > 0$. Similarly from $h_2/k_2 - a/b$ we define $h_2b - k_2a = d_2 > 0$. Solving the system,

$$k_1a - h_1b = d_1 > 0$$
$$-k_2a + h_2b = d_2 > 0,$$

routinely, we have:

$$a = \frac{\begin{vmatrix} d_1 & -h_1 \\ d_2 & h_2 \end{vmatrix}}{\begin{vmatrix} k_1 & -h_1 \\ -k_2 & h_2 \end{vmatrix}} = \frac{d_1h_2 + d_2h_1}{1},$$

and

$$b = \frac{\begin{vmatrix} k_1 & d_1 \\ -k_2 & d_2 \end{vmatrix}}{\begin{vmatrix} k_1 & -h_1 \\ -k_2 & h_2 \end{vmatrix}} = \frac{d_2k_1 + d_1k_2}{1},$$

so that

$$\frac{a}{b} = \frac{d_2h_1 + d_1h_2}{d_2k_1 + d_1k_2}$$

Not only is a/b of this form; but also any other fraction fitting between $h_1/k_1 < h_2k_2$ in subsequent Farey sequences ($N \geq b - 1$) is of this form and when reduced, must have a denominator greater or equal to b, for otherwise $h_1/k_1 < h_2/k_2$ would not be adjacent. Among these positive fractions there must be one with a least denominator; and since $d_1, d_2 > 0$, this must be the one in which $d_1 = d_2 = 1$, i.e., $a/b = (h_1 + h_2)/(k_1 + k_2)$, the mediant of the adjacent fractions in the sequence of order $b - 1$ between which a/b fits.

So we can generate all reduced fractions by forming mediants. And we have now shown that *if the Farey sequence of order $N - 1$ has the Farey properties,* *then the sequence obtained by forming the mediants (neglecting those with denominators greater than N) is indeed the Farey sequence of order N*: it contains all reduced fractions with denominators less than or equal to N in their proper order, and the difference between two adjacents is of the form $1/k_1 k_2$. Thus our induction is complete, as we can easily verify the first step.

But what interest has this game? It leads to a very interesting number theoretical result. Suppose $0 < a < b$ and a and b have no common divisor; than a/b is a reduced fraction which will appear in the Farey sequence of order b. In this sequence it will have a neighbor such that $a/b < h_2/k_2$ and $bh_2 - ak_2 = 1$.

But this is equivalent to the solution of the important Diophantine equation, $ax + by = 1$, where a and b are integers and we make up for the excess of unknowns by demanding a solution in integers. We have just shown that if *a and b are coprime, $ax + by = 1$ always has a solution in integers,* in fact, an infinite number of solutions.

Note to Chapter 4

Properties of the Farey fractions have played an important role in the so-called Hardy–Littlewood method [see G. H. Hardy (1966)]. Perhaps the most celebrated result obtained by that method is the beautiful theorem of I. M. Vinogradov (1937) that every sufficiently large odd number is expressible as the sum of three prime numbers. Using the fact that

$$\int_0^1 e^{2\pi i n x} \, dx = \begin{cases} 1, & n = 0 \\ 0, & n \neq 0 \end{cases},$$

the number of ways N is the sum of three primes is precisely $\int_0^1 \psi(x)^3 \, e^{-2\pi i N x} \, dx$ where $\psi(x) = \sum_{p \leq N} e^{2\pi i p x}$, with the sum going over primes $p \leq N$, the crucial idea of Hardy and Littlewood is to replace the integral over the interval $[0, 1]$ by a sum over the Farey fractions of order N. Thus the original concept of the integral as a sum of infinitesmals is reinstituted.

Chapter 5

Decimal Fractions

In converting common fractions into decimal fractions, confining our interest to proper fractions, we find the following cases:

$$\frac{3}{5} = 0.6, \qquad \frac{3}{40} = 0.075 \tag{I}$$

in which the decimal terminates, i.e., the final digit is zero.

$$\frac{1}{3} = 0.\overline{3}333\ldots, \qquad \frac{1}{7} = 0.\overline{142857}142857\ldots \tag{II}$$

where the decimal fraction consists of a group of digits repeated over and over, marked by overbars. This group is called the period of the decimal. In fractions of this class the period begins immediately after the decimal point.

$$\frac{5}{6} = 0.8\overline{3}33\ldots, \qquad \frac{7}{30} = 0.2\overline{3}333\ldots \tag{III}$$

which are also periodic. However, here the period does not begin immediately after the decimal point.

The first example is trivial, for we can convert the given fraction into one having a power of 10 as a denominator: $\frac{3}{5} = \frac{6}{10} = 0.6$. In fact *a terminating decimal fraction can occur only if a/b has a denominator of the form* $b = 2^\alpha 5^\beta$. Suppose $b = 2^\alpha 5^\beta$. If $\alpha > \beta$ ($\alpha < \beta$) by multiplying numerator and denominator by $5^{(\alpha-\beta)}$ $[2^{(\beta-\alpha)}]$, we can convert the given fraction into one that has 10^α $[10^\beta]$ as denominator. For example,

$$\frac{3}{40} = \frac{3}{2^3 5} = \frac{5^2}{5^2}\cdot\frac{3}{2^3\cdot 5} = \frac{75}{10^3} = \frac{75}{1000} = 0.075.$$

But if the denominator contains a factor different from 2 or 5, this is impossible — for example $7 \cdot b$ can never equal any power of 10. Thus in general we cannot transform a common fraction with an arbitrary denominator into a terminating decimal fraction.

And usually when converting a proper fraction into a decimal, we find that the long division becomes an infinite process. As an example consider $\frac{3}{41}$:

$$
\begin{array}{r}
0.\overline{07317} \\
41\overline{)3} \\
30 \\
300 \\
\underline{287} \\
130 \\
\underline{123} \\
70 \\
\underline{41} \\
290 \\
\underline{287} \\
3
\end{array}
$$

Reaching the remainder 3, we pause, for we recognize this to be identical with the dividend. If the division were to be continued, then the sequence of quotients would repeat itself, i.e., $\frac{3}{41} = 0.\overline{07317}$ is infinitely repeating — the infinite process of division is periodic.

At first glance the periodicity may seem to result from the fact that there are only ten digits which may appear in the quotient, i.e., that every decimal fraction is periodic. However, this is impossible because we can construct examples of non-periodic decimal fractions, e.g., $0.101001000100001\ldots$, where the nth 1 is followed by n zeros. Furthermore we can find decimals with very long periods — try $\frac{1}{17}$ which has sixteen digits in its period — so that some digits must reappear. The key to this process is not found in the quotients. Rather we must look to the remainders.

Suppose a/b is a reduced fraction; then there are $b - 1$ possible remainders. Zero is excluded, for the termination of the decimal fraction is identical with $b = 2^\alpha 5^\beta$ — the case completely dismissed. Thus we consider the cases where b contains primes other than 2 and 5 — and in particular we restrict our attention to case (II) where all the prime factors of b are different from 2 and 5. We will show that *if a/b is a reduced fraction and b is coprime to* 10, *then the period begins just after the decimal point*, i.e., that $(b, 10) = 1$ characterizes the fraction of case (II).

Suppose we find a remainder equal to some later remainder, i.e., $r_k = r_{k+\lambda}$. This certainly must happen, for there are only $b - 1$ possible remainders. If the assertion is true, then the first such remainder must in fact be the numerator of the fraction, for the dividend is counted as a remainder. From $r_k = r_{k+\lambda}$, we easily conclude that $r_{k+1} = r_{k+\lambda+1}$. But to show that the period must begin as early as possible, we must work backwards. We arrive at r_k and $r_{k+\lambda}$ from:

$$10r_{k-1} = q_k \cdot b + r_k, \qquad 10r_{k+\lambda-1} = q_{k+\lambda} \cdot b + r_{k+\lambda}.$$

Subtracting, we have

$$10(r_{k-1} - r_{k+\lambda-1}) = b(q_k - q_{k+\lambda}).$$

But b is coprime to 10, so that by Euclid's lemma b divides

$$r_{k-1} - r_{k+\lambda-1}, \qquad \text{i.e.,} \qquad r_{k-1} - r_{k+\lambda-1} = m \cdot b.$$

But since all $r_\nu < b$ and the absolute value of the difference of two numbers less than b must itself be less than b, i.e., $|r_{k-1} - r_{k+\lambda-1}| < b$, we have $r_{k-1} - r_{k+\lambda-1} = 0$, or $r_{k-1} = r_{k+\lambda-1}$. So that if b is coprime to 10 (the essential point used in the proof) and $r_k = r_{k+\lambda}$, then stepwise we can show that all the remainders with indices differing by λ must be equal. Hence in case (II), the periodicity must start as early as possible — we are sure that it begins immediately after the decimal.

How long is the period of such a fraction? If λ denotes the length of the period, we have found that $\lambda \le b - 1$. For the fraction, $\frac{1}{7}$, λ is 6, for $\frac{1}{17}$ it is 16, while 3 has a period of one digit. We can improve this inequality considerably. *In fact, the only residues (remainders) that can appear must be coprime to b.* If r_k is coprime to b, then rewriting

$$10r_k = q_{k+1} \cdot b + r_{k+1}$$

as

$$r_{k+1} = 10r_k - q_{k+1} \cdot b,$$

we see that, since $(10, b) = 1$ and $(r_k, b) = 1$, r_{k+1} and b can have no divisor in common. As a/b is reduced, i.e., $(a, b) = 1$, we can show stepwise that each of the remainders is coprime to b, for the preceding one is also. Thus only residues coprime to b can appear.

If b is a prime, then all the integers less than b are coprime to b. In other cases the number of coprime residues is considerably smaller. Let us introduce the customary notation: $\phi(b)$ denotes the number of residues coprime to b. In number theory, $\phi(n)$ is known as Euler's function — a function quite interesting on its own merits which we shall consider again. Here we give only a few numerical examples easily computed by inspection. First we note that $\phi(b) \le b - 1$.

$$\phi(2) = 1, \qquad \phi(3) = 2, \qquad \phi(4) = 2, \qquad \phi(5) = 4,$$
$$\phi(6) = 2, \qquad \phi(7) = 6, \qquad \phi(8) = 4, \qquad \phi(9) = 6,$$
$$\phi(10) = 4, \dots.$$

Not only is $\lambda \le b - 1$, but as we have just shown, $\lambda \le \phi(b) \le b - 1$. Moreover, not all numbers less than $\phi(b)$ can serve as length of a period. Shortly we shall prove the essential result: λ must be a divisor of $\phi(b)$. For $\frac{1}{7}$, $\lambda = \phi(7) = 6$; while for $\frac{3}{41}$, $\lambda = 5$ which is a divisor of $\phi(41) = 40$. However, if we did not know that $\frac{3}{41}$ has a period of length 5, we could only say that λ was one of the numbers 2, 4, 5, 8, 10, 20, or 40, the divisors of $\phi(41)$ — which one we could not foretell. In general the best that may be said is that λ is among the divisors of $\phi(b)$, where $\lambda = \phi(b)$ is counted as a divisor.

Let us consider in detail the example $\frac{1}{7}$

$$
\begin{array}{r}
0.\overline{142857} \\
7)\overline{1} \\
10 \\
7 \\
\hline
30 \\
28 \\
\hline
20 \\
14 \\
\hline
60 \\
56 \\
\hline
40 \\
35 \\
\hline
50 \\
49 \\
\hline
1
\end{array}
$$

$$
\begin{array}{cc}
1 & \\
5 & 3 \\
4 & 2 \\
6 &
\end{array}
$$

and look into the remainders which appear in the order: 1, 3, 2, 6, 4, 5, that can be written as a cycle. Since all the possible remainders of 7 are included in this cycle, we can write down $\frac{2}{7}$ immediately. Its sequence of remainders must begin with 2 and continue in the order: 2, 6, 4, 5, 1, 3. Hence $\frac{2}{7} = 0.285714$. Similarly $\frac{5}{7} = 0.714285$. And the other multiples of $\frac{1}{7}$ could be found by a cyclic interchange of the remainders.

Next let us examine $\frac{1}{21}$. First what is $\phi(21)$? Instead of counting the numbers coprime to 21, it is easier to count and eliminate those numbers less than or equal to 21 that have factors in common with it. Of such numbers seven are three-fold and three seven-fold; but we have counted 21 twice, so that there are nine numbers less or equal to 21 that are not coprime to it. Hence $\phi(21) = 21 - 9 = 12$. Hence we expect λ to be either 2, 3, 4, 6, or 12.

$$
\begin{array}{r}
0.\overline{047619} \\
21)\overline{1} \\
10 \\
100 \\
84 \\
\hline
160 \\
147 \\
\hline
130 \\
126 \\
\hline
40
\end{array}
$$

$$\frac{21}{190}$$
$$\frac{189}{1}$$

Here we find $\lambda = 6$ and the sequence of remainders: 1, 10, 16, 13, 4, 19. Clearly we can read off $\frac{16}{21} = 0.761907$. And similarly we could write down $\frac{10}{21} \cdot \frac{13}{21} \cdot \frac{4}{21}$, and $\frac{19}{21}$. However, as the remainder 2 does not appear in this list, we cannot write down $\frac{2}{21}$. Let us form this decimal fraction:

$$
\begin{array}{r}
0.095238 \\
21\overline{)2} \\
20 \\
200 \\
189 \\
\hline
110 \\
105 \\
\hline
50 \\
42 \\
\hline
80 \\
63 \\
\hline
170 \\
168 \\
\hline
2
\end{array}
$$

Here λ is again 6 and the new sequence of residues is: 2, 20, 11, 5, 6, 17, which we derived by considering the first possible remainder missing from the previous list. It is clear that the second scheme cannot contain any residues belonging to the first, for if it did, as the periodicity begins just after the decimal point, both sequences would be identical. Thus we have found all possible remainders, a fact to be proved later, and we suspect *that the period is indeed independent of the numerator.*

We are working with the reduced fractions a/b, i.e., $(a, b) = 1$, of the sort where b is coprime to 10; and we have found that the period begins immediately after the decimal point, i.e., $r_k = r_{k+\lambda}$ and $a = r_0 = r_\lambda$. What does this mean? Writing out the division we have:

$$10r_0 = b \cdot q_1 + r_1$$
$$10r_1 = b \cdot q_2 + r_2$$
$$\vdots \qquad \vdots$$
$$10r_{\lambda-2} = b \cdot q_{\lambda-1} + r_{\lambda-1}$$
$$10r_{\lambda-1} = b \cdot q_\lambda + r_\lambda.$$

Multiplying the first equation by $10^{\lambda-1}$, the second by $10^{\lambda-2}$, . . . , the last but one by 10, we have:

$$10^{\lambda} \cdot r_0 = b \cdot q_1 10^{\lambda-1} + r_1 \cdot 10^{\lambda-1}$$
$$10^{\lambda-1} \cdot r_1 = b \cdot q_2 10^{\lambda-2} + r_2 \cdot 10^{\lambda-2}$$
$$\vdots \qquad \vdots \qquad \qquad \vdots$$
$$10^2 \cdot r_{\lambda-2} = b \cdot q_{\lambda-1} \cdot 10 + r_{\lambda-1} 10$$
$$10 \cdot r_{\lambda-1} = b \cdot q_{\lambda} + r_{\lambda}.$$

When these are added, except for the first and last, the factors involving the r's will cancel, so we obtain:

$$10^{\lambda} r_0 = b(q_1 10^{\lambda-1} + q_2 10^{\lambda-2} + \cdots + q_{\lambda-1} 10 + q_{\lambda}) + r_{\lambda} = b \cdot Q + r_{\lambda},$$

where Q, the expression in parentheses, is the period written in the form of an integer, e.g., $\frac{1}{7} = 0.\overline{143857}$,

$$Q = 10^5 + 4 \cdot 10^4 + 2 \cdot 10^3 + 8 \cdot 10^2 + 5 \cdot 10 + 7 = 142\,857.$$

Since $r_0 = r_{\lambda} = a$, we have $10^{\lambda} a - a = b \cdot Q$ of $a(10^{\lambda} - 1) = b \cdot Q$. And as a and b are coprime, b divides $10^{\lambda} - 1$. This tells us two things:

1. *there is a λth power of 10 which diminished by 1 is divisible by b*;
2. *λ is independent of a*, for clearly λ is the smallest such exponent that we could choose such that $10^{\lambda} - 1$ is divisible by b.

Thus the results we obtained in the case $b = 21$ are not fortuitous — necessarily the period of $\frac{1}{21}$ has the same length as that of $\frac{2}{21}$. On closer examination of the sequences of residues:

$$\frac{1}{21} \quad \text{yielding} \quad 1, 10, 16, 13, 4, 19\,,$$

$$\frac{2}{21} \quad \text{yielding} \quad 2, 20, 11, 5, 8, 17\,,$$

we see that there is indeed a relation between them, namely those of the second sequence are twice the corresponding one of the first, diminished by 21, if necessary, to produce a residue less than 21, e.g., from 16 we have $2 \cdot 16 = 32$ and $32 - 21 = 11$, the third residue of the second sequence, etc.

Now we can easily show that *the length of the period, λ, must be a divisor of $\phi(b)$*. Dividing b into 1, we obtain a sequence of residues: $1, r_1, r_2, \ldots, r_{\lambda-1}$, all of which are coprime to b. If the number of these does not exhaust $\phi(b)$, then we can choose some other residue coprime to b, say r_0', and dividing this by b, obtain a new sequence of residues: $r_0', r_1', \ldots, r_{\lambda-1}'$, which must be of the same length as the first. All members of this second scheme must be different from those of the previous one. For if not, since one residue entirely determines the whole sequence of residues, if a remainder of both sequences were identical, then the second would merely be the first all over again. But r_0' is not contained in the first.

Hence we have found 2λ residues which are all different. These may or may not exhaust $\phi(b)$. If not, we find a third set of residues, λ in number, different from the preceding, etc. Since $\phi(b)$ is finite, eventually we must exhaust it by forming new and entirely different sequences of residues, which can only occur in sets containing λ each. Thus $\phi(b)$ must be a multiple of λ.

$$\phi(b) = k \cdot \lambda(b),$$

where we write $\lambda(b)$ to emphasize the dependence of λ on b. As examples we have:

$$\phi(21) = 12, \quad \lambda(21) = 6, \quad k = 2,$$
$$\phi(41) = 40, \quad \lambda(41) = 5, \quad k = 8,$$
$$\phi(\ 7) = \ 6, \quad \lambda(\ 7) = 6, \quad k = 1.$$

Let us return to the first conclusion we drew from the relation $a(10^\lambda - 1) = b \cdot Q$, namely: $10^\lambda - 1$ is divisible by b. To derive a more abstract result from this, we note that $x^k - 1$ is divisible algebraically by $x - 1$, i.e., $(x^k - 1) = (x - 1)(x^{k-1} + x^{k-2} + \cdots + 1)$. Replacing x by 10^λ, we have: $10^{\lambda k} - 1 = (10^\lambda - 1)X$, where X is a polynomial in 10^λ. Since this is an algebraic identity, we may choose such a k that $10^{k\lambda} = 10^{\phi(b)}$. And as a plays no role in determining λ or k, putting $a = 1$, we have $10^\lambda - 1 = bQ$. This gives $10^{\phi(b)} - 1 = bQX$. Whence we conclude: $10^{\phi(b)} - 1$ *is divisible by* b, *provided* $(10, b) = 1$ — a theorem on integers, one having nothing to do with fractions.

This is the famous Fermat–Euler theorem. If p is a prime, $\phi(p) = p - 1$, the formula reads $10^{p-1} - 1$ is divisible by P, provided $p \neq 2, 5$. This theorem was known to Fermat, one of the very greatest mathematicians of the 17th century, if not of all time — a jurist whose contributions to mathematics, his hobby, made him immortal, whereas his jurisprudence is forgotten. Euler put the theorem in its most general form, which we shall shortly produce, but the essential idea is due to Fermat. Let us make a few examples: Taking p as small as possible, i.e., 3, we have $10^2 - 1 = 99$ is divisible by 3; or for $p = 7$, we have $10^6 - 1 = 999\,999$ is divisible by 7 (recall $10^6 - 1 = 7.142859$).

Here we have the special number 10 in our formula, as a result of the fact that we used the decimal system to write our fractions. However, the whole argument could be reproduced using a g-adic number system, in which $\alpha\beta\gamma$ means $\alpha g^2 + \beta g + \gamma$ and, $\alpha\beta\gamma$ means $\alpha g^{-1} + \beta g^{-2} + \gamma g^{-3}$, defining periodic decimal fractions in the same way. (To the mathematician the discussion of these different number systems isn't very interesting, for they are merely notations that have nothing to do with the fundamental nature of numbers. The one that we use is merely a linguistic heritage — probably a biological accident in that we have 10 fingers — although our language does have remnants of other systems of notation, notably dozen, score, and gross.)

Thus we may replace 10 by any number coprime to b, and obtain: $g^{\phi(b)} - 1$ *is divisible by* b, *provided that* g *and* b *are coprime,* the Fermat–Euler theorem in its most general formulation. This theorem we see provides the background for the systematic study of decimal fractions.

Let us make an example: $2^{p-1} - 1$ is divisible by p, where p is a prime greater than 2.

$p = 3$:	$2^2 - 1 = 4 - 1 = 3$	divisible by 3;
$p = 5$:	$2^4 - 1 = 16 - 1 = 15$	divisible by 5;
$p = 7$:	$2^6 - 1 = 64 - 1 = 63$	divisible by 7;
$p = 11$:	$2^{10} - 1 = 1024 - 1 = 1023$	divisible by 11 , etc.

To satisfy curiosity we state here Fermat's Last Theorem. We could show that there are infinitely many integers, the so-called Pythagorean numbers, that satisfy the equation $a^2 + b^2 = c^2$, e.g., $3^2 + 4^2 = 5^2$, $5^2 + 12^2 = 13^2$, etc. Fermat claimed there were no integers satisfying $a^n + b^n = c^n$ for $n > 2$. We do now know that this is true for many n, but it is still not proved in full generality. In part the interest of the theorem lies in the provocative way in which it was first stated. Fermat wrote the assertion on his copy of Diophantus together with the remark, "...I have discovered a truly marvellous demonstration which this margin is too narrow to contain." However, the importance of the theorem lies not in its content, but in the mathematics developed in the attempts to prove it — the efforts to do so in the 19th Century yielded the new field of algebraic number theory and the notion of ideal numbers developed first by Kummer.

To close this chapter, let us set the converse problem: *Given a periodic decimal fraction, to what common fraction does it belong?* We have $a(10^\lambda - 1) = bQ$, where Q is the period of the decimal fraction read as an integer. Thus $a/b = Q/(10^\lambda - 1)$. For example, to what common fraction does $0.\overline{09}$ belong? Here $Q = 9$, $\lambda = 2$, so that $a/b = 9/(10^\lambda - 1) = 9/99 = 1/11$.

Finally we note that the third case, which has not been treated here, that in which the period of the decimal fraction does not begin immediately after the decimal point, is the mixed case in which b has a divisor in common with 10.

Note to Chapter 5

In recent years some remarkable applications of the Fermat–Euler theorem have come to light. It was a surprise to many people to learn that a procedure was developed whereby a secret message could be encoded and the person encoding the message would not be able to reverse the process and decode the message. The procedure is as follows. Let N be a very large number which has at least two prime factors. Assume for simplicity that $N = p_1 p_2$, the product of two large primes. Then $\phi(N) = (p_1 - 1)(p_2 - 1)$. Now, if p_1 and p_2 are very large, then it could take a computer hundreds of years to factor N, and hence a person knowing only N could never really determine $\phi(N)$. Let E and D be two integers satisfying $ED = \phi(N) + 1$. The person encoding the message is given the numbers N and E. As shown above, he cannot know D. Let us assume he has a message M (given in the form of a large number, say). Then he encodes the message by computing M^E, dividing this by N, and computing the remainder R. In other words, $M^E - R$ is exactly divisible by N. Then R will be the encoded message. It is not possible to decode this message without knowing D. To decode the message, one simply takes R^D and computes the remainder after dividing by N. This is based on the fact that $R^D \equiv M^{ED} \equiv M^{\phi(N)+1} \equiv M \pmod{N}$ by Euler's theorem. As a simple example, take $N = 33, \phi(N) = 20, E = 7, D = 3, M = 2$. Since $2^7 \equiv 29 \pmod{33}$, the encoded message is $R = 29$. To decode the message, note that $29^3 \equiv 2 \pmod{33}$.

Chapter 6

A Theorem of Formal Logic

The formula discussed in this chapter is generally ascribed to the British mathematician, Sylvester, who taught for some years in this country, both at the University of Virginia and at Johns Hopkins. However, the theorem must certainly have appeared earlier in investigations on formal logic.

We come naturally to the ideal from the consideration of the calculation of Euler's function, $\phi(n)$. How did we compute $\phi(21)$? As we wanted to retain only the numbers, $a \leq 21$, for which $(21, a) = 1$, we counted and eliminated those not having this property, i.e., $\phi(21) = 21 - \frac{21}{3} - \frac{21}{7} + \frac{21}{21}$, where the last term was added to take care of the fact that we had subtracted one term too many.

Now let us do this a bit more generally; what is $\phi(p^\alpha q^\beta)$, where p and q are prime numbers? First of all, there are $p^\alpha q^\beta$ remainders, including $p^\alpha q^\beta$ itself. Some of these are multiples of p, in fact $p^\alpha q^\beta/p$ of them, so this number must be subtracted from the total number of remainders. Similarly there are $p^\alpha q^\beta/q$ numbers that are multiples of q. But among the multiples of p there occur the multiples of pq and these also are found among the multiples of q, so that we have subtracted these twice. Hence we must add $p^\alpha q^\beta/pq$. Thus

$$\phi(p^\alpha q^\beta) = p^\alpha q^\beta - \frac{p^\alpha q^\beta}{p} - \frac{p^\alpha q^\beta}{q} + \frac{p^\alpha q^\beta}{pq}$$

$$= p^\alpha q^\beta \left(1 - \frac{1}{p} - \frac{1}{q} + \frac{1}{pq} \right)$$

$$= p^\alpha q^\beta \left(1 - \frac{1}{p} \right) \left(1 - \frac{1}{q} \right),$$

a formula to which we will return. However, at the moment this serves as a concrete introduction to a far more general ideal — to the example of completely abstract logical reasoning which is the meat of this chapter.

Let us suppose that we have a set of objects S, N in number, which have different properties, e.g., red, green, round, and rough, some of which perhaps are mutually exclusive, some perhaps not. Let us denote these attributes which may be predicated of the objects of S by Greek letters, α, β, γ, etc. Let N_α be the number of objects in S possessing the attribute α, $N_{\alpha\beta}$ be the number of objects in S possessing both the attributes α and β, etc.; and let $\overline{N}$ be the number of objects in S that possess none of a given list of attributes: α, β, γ, ...

The problem is to find the number of objects in S that possess neither α, nor β, nor γ, ... We shall show that the solution is

$$\overline{N} = N - N_\alpha - N_\beta - N_\gamma - \cdots$$
$$+ N_{\alpha\beta} + N_{\alpha\gamma} + \cdots + N_{\beta\gamma} + \cdots$$
$$- N_{\alpha\beta\gamma} - N_{\alpha\beta\delta} - \cdots$$
$$+ N_{\alpha\beta\gamma\delta} + \cdots ,$$

where the formula is understood to be continued in the foregoing pattern with minus signs preceding the number of objects possessing an odd number of attributes and the plus sign preceding the number of objects possessing an even number of attributes until all possible combinations of attributes have been exhausted.

In general let us denote the negation of an attribute by a bar over the attribute negated, i.e., $N_{\overline{\alpha}}$ is the number of objects that do not possess the attribute α, that are "non-α."

First let us examine the formula in the simplest case. Suppose we wish to know the number of objects in S that do not have an attribute α. Since we are concerned with a single property, we can think of all members of S as being either α or non-α, so that $\overline{N} = N_{\overline{\alpha}} = N - N_\alpha$, which is our formula for a single attribute.

Next we consider two properties α and β which may be attributed to the members of S and ask how many objects there are in S that have neither the attribute α nor the attribute β. This number we denote by $\overline{N} = N_{\overline{\alpha},\overline{\beta}}$. Now let us introduce an auxiliary property A such that an object will be said to have the attribute A if it possesses either the property α or the property β. This we write more briefly as $A = (\alpha \text{ or } \beta) = (\alpha \vee \beta)$, where the symbol "$\vee$" is understood to mean "or" in the sense of "and/or" as used in joint bank accounts. For example, let S consist of a bag of marbles possessing the properties α = red, β = green, γ = uncolored, so that the property $A = (\alpha \vee \beta)$ is being colored. Thus we reduce the question for two properties to the preceding case with a single property, and we write

$$\overline{N} = N_{\overline{\alpha},\overline{\beta}} = N_{\overline{A}} = N - N_A .$$

Now we must find N_A. This is certainly related to the sum $N_\alpha + N_\beta$ and in fact is equal to it in the case when α and β are mutually exclusive properties. However, α and β, as we have admitted from the beginning, may not exclude one another; if α = red and β = rough, then we can have marbles that are both red and rough—a situation we represent schematically in Fig.1. Thus we have counted

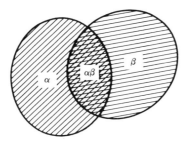

Figure 1

the items possessing both α and β once too often, so, to compensate, we subtract $N_{\alpha\beta}$. Hence $N_A = N_\alpha + N_\beta - N_{\alpha\beta}$, and $\overline{N} = N_{\overline{\alpha},\overline{\beta}} = N - N_\alpha - N_\beta - N_{\alpha\beta}$, our formula for two attributes.

The reduction of the case of two attributes to that of one is a good start toward the proof of the formula in general by induction. Suppose the formula is correct for a given number of properties, α, β, γ, $\ldots$ and that ξ is a new attribute applicable to the elements of S. We want to calculate the number of elements that are neither α, nor β, nor γ, $\ldots$, nor ξ, i.e., $N_{\overline{\xi},\overline{\alpha},\overline{\beta},\overline{\gamma}, \ldots}$. In S there is a certain set of objects that have the property ξ, N_ξ in number. To this set, by hypothesis, we can apply the formula to determine the number of elements that are non-α, non-β, non-γ, $\ldots$, i.e., $N_{\xi,\overline{\alpha},\overline{\beta},\overline{\gamma}, \ldots}$. To obtain this number, all we need to do is replace the N of the formula by N_ξ:

$$
\begin{aligned}
N_{\xi,\overline{\alpha},\overline{\beta},\overline{\gamma}, \ldots} = \; & N_\xi - N_{\xi\alpha} - N_{\xi\beta} - N_{\xi\gamma} - \cdots \\
& + N_{\xi\alpha\beta} + N_{\xi\alpha\gamma} + N_{\xi\alpha\delta} + \cdots \\
& - N_{\xi\alpha\beta\gamma} - N_{\xi\alpha\beta\delta} - \cdots \\
& + N_{\xi\alpha\beta\gamma\delta} + \cdots .
\end{aligned}
$$

This set having the property ξ but none of the properties α, β, γ, $\ldots$ is but a part of the larger set, numbering $N_{\overline{\alpha},\overline{\beta},\overline{\gamma}, \ldots}$, which is non-α, non-β, non-γ, $\ldots$. To fill out the latter, we must add to the former the set that does not have the property ξ and lacks as well the attributes α, β, γ, $\ldots$, i.e., the set we are seeking to enumerate. Thus:

$$
N_{\overline{\xi},\overline{\alpha},\overline{\beta},\overline{\gamma}, \ldots} + N_{\xi,\overline{\alpha},\overline{\beta}, \ldots} = N_{\overline{\alpha},\overline{\beta},\overline{\gamma}, \ldots} .
$$

But we take for granted that we can apply the formula to determine

$$
\begin{aligned}
N_{\overline{\alpha},\overline{\beta},\overline{\gamma}, \ldots} = \; & N - N_\alpha - N_\beta - N_\gamma - \cdots \\
& + N_{\alpha\beta} + N_{\alpha\gamma} + \cdots \\
& - N_{\alpha\beta\gamma} \ldots .
\end{aligned}
$$

Hence we have:

$$N_{\bar{\xi},\bar{\alpha},\bar{\beta},\bar{\gamma},\,\ldots} = N_{\bar{\alpha},\bar{\beta},\bar{\gamma},\,\ldots} - N_{\xi,\bar{\alpha},\bar{\beta},\bar{\gamma},\,\ldots}$$
$$= \{N - N_\alpha - N_\beta - N_\gamma - \cdots$$
$$+ N_{\alpha\beta} + N_{\alpha\gamma} + \cdots$$
$$- N_{\alpha\beta\gamma} - \cdots \}$$
$$- \{N_\xi - N_{\xi\alpha} - N_{\xi\beta} - \cdots$$
$$+ N_{\xi\alpha\beta} + N_{\xi\alpha\gamma} + \cdots \}$$
$$= N - N_\alpha - N_\beta - N_\gamma N_{-\delta} \cdots - N_\xi$$
$$+ N_{\alpha\beta} + N_{\alpha\gamma} + \cdots + N_{\xi\alpha} + N_{\xi\beta} \cdots$$
$$- N_{\alpha\beta\gamma} - \cdots - N_{\xi\alpha\beta} - \cdots$$
$$+ N_{\alpha\beta\gamma\delta} \cdots + N_{\xi\alpha\beta\gamma} + \cdots .$$

Thus we see that, if the formula is valid for a given number of properties, it emerges with the same form when another property is added to the list. Hence the induction is complete.

A shorter proof of this formula involving the binomial coefficients is frequently given [see Pólya, Szegö (1964)]. However, as we propose to extend the result to classes of mathematical objects even wider than those subject to mere counting, we find the general proof more suitable.

But first let us reconsider the computation of $\phi(m)$. Let m be decomposed into prime factors, so that $m = p^a q^b r^c \ldots$, where $p, q, r, \ldots$ are all different primes. (Recall that we proved this to be unique.) What do we want to find? $\phi(m)$ is the number of integers of the sequence $1, 2, 3, \ldots , m$ which are relatively prime to m. A number not prime to m will have a divisor in common with it. Thus we have to count the numbers less than p that have the attribute of being neither divisible by p (the property α), nor by q (the property β), nor by r (the property γ), $\ldots$. Hence $\phi(m)$ is merely a way of writing $N_{\bar{\alpha},\bar{\beta},\bar{\delta},\,\ldots}$ in a particular application of our formula. If we denote the number of multiples of p less than or equal to m by m_p, etc., we can write:

$$\phi(m) = m - m_p - m_q - m_r - \cdots$$
$$+ m_{pq} + m_{pr} + \cdots$$
$$- m_{pqr} - \cdots ,$$

which will expire as soon as we have exhausted the supply of primes available in m. In this case we can easily compute the numbers m_p, etc., so that

$$\phi(m) = m - \frac{m}{p} - \frac{m}{q} - \frac{m}{r} - \cdots + \frac{m}{pq} + \frac{m}{pr} + \cdots - \frac{m}{pqr} - \cdots$$
$$= m\left(1 - \frac{1}{p} - \frac{1}{q} - \frac{1}{r} - \cdots + \frac{1}{pq} + \frac{1}{pr} + \cdots - \frac{1}{pqr} - \cdots\right)$$
$$= m\left(1 - \frac{1}{p}\right)\left(1 - \frac{1}{q}\right)\left(1 - \frac{1}{r}\right)\cdots .$$

This we can write more concisely as

$$\phi(m) = m \prod_{p|m} \left(1 - \frac{1}{p}\right),$$

where the symbol $p|m$ indicates that the product is to be taken over all prime factors of m.

As an example let us compute

$$\phi(60) = 60 \prod_{p|60} \left(1 - \frac{1}{p}\right) = 60\left(1 - \frac{1}{2}\right)\left(1 - \frac{1}{3}\right)\left(1 - \frac{1}{5}\right),$$

which must always be an integer. Or,

$$\phi(60) = 60 \cdot \frac{1}{2} \cdot \frac{2}{3} \cdot \frac{4}{5} = 16,$$

a fairly low number for m as large a number as 60. Furthermore, since a prime has only one prime divisor, itself, we find that $\phi(p) = p(1 - 1/p) = p - 1$, so that the formula checks in this well known case.

If this were all the Sylvester formula were good for, we should not have emphasized it. But there is indeed more.

So far we have spoken only of finite sets. But suppose we have an *infinite set*: How can we compare it with other sets of similar nature, how can we "measure" such a set? A finite set is most simply measured by counting. But there are many other ways of measuring things — length and area are common measures that may be applicable to certain infinite sets.

In order to show that the proof of the Sylvester formula is applicable to such sets, we need the notion of an *additive measure*. This is the only thing that was essential in the proof we gave. It is such a simple notion that we usually take it for granted. When from two finite and disjunct (having no elements in common) sets A and B, we form the combined set $A + B$, consisting of the elements belonging to A or to B, we scarcely notice that the number of items in $A + B$ is equal to the sum of the number of items in A and in B separately. But if we think of this counting as a measure, denoting it by m, we can specifically note this fact succinctly in the equation: $m(A + B) = m(A) + m(B)$ and briefly describe it by saying that the measure is *additive*. In the same way we see that length is an additive measure: for if we have two infinite sets consisting of the points on the line segments s and r, to which length is an applicable measure, then we know that the length of $s + r$ is equal to the sum of the individual lengths, i.e., $m(s + r) = m(s) + m(r)$ (see Fig. 2). And finally, area as a measure is also additive: $m(A + B) = m(A) + m(B)$. If this property is lacking in our measure, then either we throw out the problem as being unsolvable with the traditional tools, or we construct a more involved theory of measure as Lebesque has done.

Now our proof of Sylvester's formula still holds for these more general sets, for the fact that the measure was additive was the only important property that we used. This we used fully. Let S be a set having an additive measure $m(S)$, more briefly mS. Then if S_α is a subset of S having a property α, we certainly have

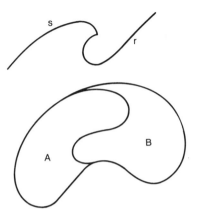

Figure 2

$S = S_\alpha + S_{\bar{\alpha}}$. And if the measure of S is also applicable to S_α and $S_{\bar{\alpha}}$, we have $mS = mS_\alpha + mS_{\bar{\alpha}}$ which we write as:

$$mS_{\bar{\alpha}} = mS - mS_\alpha.$$

We recognize this as the first case of Sylvester's formula applied to these more general sets and that the deduction is completely analogous to the one we made previously. Putting together the two sets $S_{\xi,\bar{\alpha},\bar{\beta},\bar{\gamma},\,\ldots}$ and $S_{\bar{\xi},\bar{\alpha},\bar{\beta},\bar{\gamma},\,\ldots}$ to form the set $S_{\bar{\alpha},\bar{\beta},\bar{\gamma},\,\ldots}$ is the crucial point of the induction. Here again the whole thing rests on the rock of additivity. If the measure satisfies this condition, then the proof is valid. And we can write:

$$S_{\bar{\xi},\bar{\alpha},\bar{\beta},\bar{\gamma},\,\ldots} = S_{\bar{\alpha},\bar{\beta},\bar{\gamma},\,\ldots} - S_{\xi,\bar{\alpha},\bar{\beta},\bar{\gamma},\,\ldots}\,,$$

from which the induction follows.

This cumbersome notation has served to make clear the analogy. Now let us replace it by one briefer and more suitable. Let the Latin capitals, A, B, C, $\ldots$ denote subsets included in the set S, a relation that we symbolize as $A \subset S, B \subset S, \ldots$. Clearly these purely arbitrary subsets can play the same role in Sylvester's formula as S_α, $S_{\bar{\alpha}}$, $\ldots$ which we have thought of as being defined by some "attribute." We call two subsets A and B disjunct if they have no elements in common and note this by writing $A \cdot B = 0$, where by $A \cdot B$, or AB, we understand that set which consists of the elements included both in A and in B (see Fig. 3). Using this notation, we can express concisely what we mean by an additive measure: If $A \cdot B = 0$ and $m(A + B) = mA + mB$, the measure m is additive. The iteration of this single notion enabled us to prove the rule, which we now write more simply:

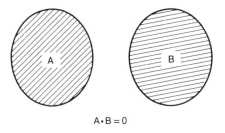

$$A \cdot B = 0$$

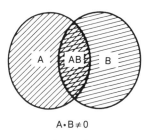

$$A \cdot B \neq 0$$

Figure 3

$$m(S_{\overline{ABC}} \ldots) = mS - mA - mB - mC - \cdots$$
$$+ mAB + mAC + \cdots + mBC + \cdots$$
$$- mABC - mABD - \cdots$$
$$+ mABCD + \cdots$$
$$\cdots ,$$

where $A, B, C, D, \ldots \subset S$ and by $S_{\overline{ABCD}} \ldots \subset S$ we mean that part of S which has nothing in common with $A, B, C, \ldots$, e.g., if in Fig. 4, S is the whole interior of the heavy boundary, $S_{\overline{AB}}$ is the unshaded part of the interior. Figure 4 also serves as an example of

$$m(S_{\overline{AB}}) = mS - mA - mB + mAB .$$

Application to Probability

The notions connected with this theorem form the basis for recent advances in the theory of probability. There the essential problem is to understand what we mean by *probability*.

This word is often confused with relative frequency. For example, when we throw a die and count the number of times an ace appears in a long sequence of

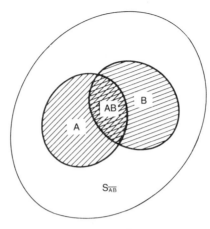

Figure 4

throws, we are inclined to say that the relative frequency approaches $\frac{1}{6}$, i.e., the number of successes (aces) $\rightarrow \frac{1}{6}$, which we know to be the probability of such an event. This statement has no meaning because the notion of the convergence of a sequence in the precise mathematical sense is only applicable to an infinite sequence — and certainly we can never obtain an infinite sequence by throwing a die.

As a matter of fact the relative frequency that we know must have something to do with probability is a question in the practical application of the theory which does not concern us here. Even though we should find a loaded die that has shown an ace on every known throw, we can never be sure that a 6 will not appear on the next cast. Physics has long thrived on the notion that improbable events do not occur. This is a reasonable, practical rule; but one that has no place in the theory of probability which is a mathematical discipline capable of precision.

The practical application of geometry provides similar examples of the gulf between the theory and the applications. Apply the geometry of a circle to a grinding wheel. In so far as the grinding wheel is a circle it must have properties which the geometry of the circle anticipates. But put the wheel under a magnifying glass. The edge is marked by pits and scorings. Such a thing is not a circle as we understand it in geometry. Again, in so far as light travels in straight lines, as theodolites measure angles and have pivots that are points, we may speak of the angles of a large triangle in a survey. In so far as these things are true, the Euclidean geometry is a reasonable model of finite space. But look at the atmosphere whose density changes from point to point, so that light, being in small measure refracted everywhere, does not travel in a geometric straight line. Yet we find it useful to study geometry abstractly as a precise discipline — in spite of this gulf between the theory and the applications. In the theory of probability the same situation arises, but with an added difficulty — the applications have been confused with the theory.

Probability should be looked upon quite abstractly as an additive measure. We are concerned with sets of things which we call *events*. The language is misleading, for one cannot speak of the probability of a single event. "What is the probability that Napoleon lost Waterloo?" is sheer nonsense. A single, specific, historic event is fully determined and has no probability. When we speak of probability, we must always speak of classes of events. First we consider a universal event, U, e.g., all the throws of a die, and we stipulate that the probability (the measure) of U is 1, which we write $p(U) = 1$. To this universal event there belong subclasses, say $A \subset U$, which are called events. For example, the throwing of a die gives us six events, $A_1, A_2, A_3, \ldots A_6$, such that $A_j \cdot A_k = 0$ (what does this mean?) and $A_1 + A_2 + \cdots + A_6 = U$. Since probability is an additive measure, we have:

$$p(A_1) + p(A_2) + \cdots + p(A_6) = p(U) = 1,$$

which is the Sylvester formula in the special case where the measure of the set $S_{\bar{A}_1 \bar{A}_2 \ldots \bar{A}_6}$ is zero.

Books on the elementary theory of probability delight in combinatorial analysis, which has little to do with this subject. It is purely accidental that this elaborate computational machinery forms a small chapter in the theory of probability which is primarily concerned with the deriving of complicated probabilities from elementary ones.

We have already set forth a first axiom of the theory of probability, namely $m(U) = p(U) = 1$. And we have indicated as a second that probability is an additive measure. We may rewrite our former definition, $m(A) + m(B) = m(A + B)$, provided $A \cdot B = 0$, in the notation now useful, as:

$$p(A) + p(B) = p(A + B),$$

provided $A \cdot B = 0$, where in the language of probability $A \cdot B = 0$, formerly read as "the sets are disjunct," now is read as "the events A and B are mutually exclusive." This axiom is already an example of the computation of complicated probabilities from simpler ones. With these two axioms and the fact that the measure we call probability is always non-negative, we can easily show that $0 \leq p(A) \leq p(U) = 1$, provided $A \subset U$, and that if $A \subset B$, then $p(A) \leq p(B)$.

In the applications, how we obtain our elementary probabilities is a matter of good luck in so far as the theory is concerned. Yet there must be some rules that can help us solve this problem. In fact there are certain cases where we can compute these elementary probabilities. How?

There is a principle, very rarely enunciated, which is related to the more frequent statement: "Equally likely events have equal probabilities." This is something of a vicious circle; yet behind it there lies a worthwhile notion that we may express precisely. We postulate that events have the same probability or they differ only by name and by no physical difference, by no essential of the conditions. For example in the casting of a die, the events $A_1, A_2, \ldots, A_6$ differ only in that we give distinct names to the sides of the cube; and collectively they fill out the universal event, U. Thus on the basis of this postulate we have:

$$p(A_1) + p(A_2) + \cdots + p(A_6) = 6p(A_1) = p(U) = 1,$$

so that

$$p(A_1) = p(A_2) > \cdots = p(A_6) = \tfrac{1}{6}.$$

This is what we believe to be so; if the probability of throwing an ace were not $\tfrac{1}{6}$, we should be strongly tempted to reject the theory. That this in fact cannot materialize is of no consequence, for we do think of a die as being made so that the sides differ only by name and by nothing in the structure of the die. Of course whether an actual die fulfills this condition or not is a question of application and does not interest us in the theory itself.

Thus we assume *a priori* that classes of events which are invariant with respect to names have equal probabilities.

We might go deeper into the theory of probability; but we content ourselves with the exposition of these four postulates:

1. $p(U) = 1$,
2. $p(A \vee B) = p(A) + p(B)$, provided $A \cdot B = 0$,
3. $p(A) > 0$,
4. events invariant with respect to names have equal probabilities, and conclude with an example of the application of Sylvester's formula.

Suppose, given the events A, B, C, and $D \subset U$, which may or may not be mutually exclusive, we seek the probability that non-A, non-B, non-C, and non-D occur simultaneously. As probability is an additive measure, Sylvester's formula is applicable, so that:

$$
\begin{aligned}
p(\overline{ABCD}) = {} & p(U) - p(A) - p(B) - p(C) - p(D) \\
& + p(AB) + p(AC) + p(AD) + p(BC) + p(BD) + p(CD) \\
& - p(ABC) - p(ABD) - p(ACD) - p(BCD) \\
& + p(ABCD).
\end{aligned}
$$

This result may be generalized to any finite number of events and is an excellent example of a theorem in the theory of probability, of the derivation of complicated probabilities from elementary ones [see, e.g., J. V. Uspensky (1937)].

Chapter 7

On the Approximation of Irrational Numbers by Rational Numbers

The approximation of irrational numbers by rationals is fundamental to practical computation. As the periodic decimals are identical with the rational numbers, any non-periodic infinite decimal represents an irrational number. When we write $\pi = 3.14159\ldots$, this irrational number is being approximated by a rational having a power of 10 as denominator. However, some irrationals can be surprisingly well approximated by very simple rational numbers, e.g., $\pi \approx \frac{22}{7} = 3.\overline{142857}$. Such observations have led to very interesting systematic discussions of the questions: How good an approximation can we obtain for a given irrational number, ω? And how easily can we approximate such a number? That is, how large a denominator must we choose so that $|\omega - m/n|$ is less than a prescribed number?

Suppose n is given, then trivially we can choose m so that $|\omega - m/n| < 1/n$ (see Fig. 5). Let us lay out our numbers on a line and mark off intervals of length $1/n$. Since ω is irrational, it will fall in the interior of one of these intervals whose end points (m_1/n and m_2/n) will satisfy the inequality $|\omega - m/n| < 1/n$. Thus *we can approximate an irrational as closely as we choose simply by taking n sufficiently large.*

Figure 5

However we can do better than this: we will show that a rational m/n can always be found such that $|\omega - m/n| < 1/n^2$—that is, among the approximations satisfying $|\omega - m/n| < 1/n$ there are some that are remarkably better than others in that they also satisfy the inequality $|\omega - m/n| < 1/n^2$.

In showing this, we will employ an argument, frequently used by Dirichlet, which is familiarly known as the *pigeon-hole principle: If N + 1 objects are distributed into N boxes, then at least one box must contain more than one object.* This simple remark will have far-reaching consequences.

Let us consider the numbers $x\omega$, where $x = 1, 2, 3, \ldots, N$. These are all irrational: For suppose $x\omega = z$ were rational, then $\omega = z/x$ would also be rational. This is impossible; so that if ω is irrational, so is any multiple of it. Without loss of generality we may assume $\omega > 0$. From $x\omega$ let us subtract the largest integer contained in $x\omega$ (this is written $[x\omega]$, e.g., $[\pi] = 3$, $[\sqrt{2}] = 1$). By forming $x\omega - [x\omega] = x\omega - y$, where y is an integer, we obtain the N numbers, such that $0 < x\omega - y_x < 1$, listed below.

$$1 \cdot \omega - y_1$$
$$2 \cdot \omega - y_2$$
$$3 \cdot \omega - y_3$$
$$N \cdot \omega - y_N$$

For pigeon-holes we take the N interval of length $1/N$ contained in the interval $0 \le x \le 1$. As $x\omega - [x\omega]$ is irrational, none of these numbers can fall at an end point of an interval.

Now we make a case distinction:[1]

1. Suppose the first pigeon-hole is occupied. Then this number satisfies the inequality $0 < x\omega - y < 1/N$, or $0 < \omega - y/x < 1/N^x \le 1/x^2$ as $0 < x \le N$. Thus in this case we have shown that a rational approximation to ω exists such that $|\omega - y/x| < 1/x^2$.

2. If the first pigeon-hole is empty, then we have N objects distributed into $N - 1$ pigeon-holes. Of course some may be empty; but at least one of them must have two numbers in it, i.e., we have:

$$\frac{j}{N} < x_a\omega - y_a < \frac{j+1}{N}$$

and

$$\frac{j}{N} < x_b\omega - y_b < \frac{j+1}{N},$$

where x_a and $x_b \le N$. Multiplying the second inequality by -1, these become:

$$\frac{j}{N} < x_a\omega - y_a < \frac{j+1}{N}$$
$$-\frac{j+1}{N} < -x_b\omega + y_b < -\frac{j}{N}.$$

And adding, we have

$$-\frac{1}{N} < (x_a - x_b)\omega - (y_a - y_b) < \frac{1}{N}.$$

[1]The proof can be arranged to avoid this.

This suffices to make the proof. Without loss of generality we may assume $x_a > x_b$, so that the differences are positive. Let $(x_a - x_b) = X < N$, as $x_b < x_a \leq N$, and $Y = (y_a - y_b)$.

Now our inequality may be written as:

$$-\frac{1}{N} < X\omega - Y < \frac{1}{N}, \quad \text{or} \quad |X\omega - Y| < \frac{1}{N}.$$

And as $X > 0$, this yields:

$$\left| \omega - \frac{Y}{X} \right| < \frac{1}{XN} < \frac{1}{X^2}.$$

Thus in both cases we have found a rational approximation to ω which differs from it by less than the reciprocal square of the denominator. More precisely we have the theorem: *If ω is irrational, there always exists Y/X, such that $1 \leq X \leq N$, satisfying the inequality $|\omega - Y/X| < 1/NX$.*

Is there only one of these better approximations to ω? No! In fact there are infinitely many, just as there are infinitely many n satisfying $|\omega - m/n| < 1/n$. How does this happen? Given N_1 we can find an approximation $|\omega - Y_1/X_1| < 1/X_1^2$. Choosing N_2 such that $|\omega - Y_1/X_1| = \delta > 1/N_2$, then, as our theorem asserts, we can find Y_2/X_2, such that $|\omega - Y_2/X_2| < 1/N_2X_2 \leq 1/X_2^2$, an approximation that is certainly different from the first one. Clearly we can continue to find new approximations, each different from the preceding ones, so that the inequality $|\omega - Y/X| < 1/X^2$ is satisfied infinitely often. This means that among the rational approximations, m/n, to ω there are infinitely many that are better than one would suspect from examining their denominators, that differ from ω by less than $1/n^2$. However, we have only shown that given N, we can find an $X \leq N$ such that $|\omega - Y/X| < 1/NX \leq 1/X^2$. Thus not every denominator will do; only rather clever choices may serve as denominators of such fractions. The approximation of π by $22/7$ is an outstanding example: $|\pi - 22/7| = |3.14159 \ldots - 3.\overline{142857}| < 0.0013$ which is very much better than $1/7^2 = 1/49 < 0.0205$.

Since we can find infinitely many approximations to ω of the order $1/N$ and $1/N^2$, it is natural to ask can we find infinitely many approximations of the order $1/N^3$, or in general to what order can we approximate an irrational. Liouville (1851), the founder of *Le Journal des Mathématiques Pures et Appliquées,* whose name is connected with the theory of functions (especially elliptic functions) as well as with the theory of numbers, answered these questions. He showed that in general it is impossible to find approximations to an irrational number of the order of $1/N^3$. He showed that some irrational numbers resist an inequality of the form $|\omega - y/x| < 1/x^k, k > 2$.

These reluctant numbers are the *algebraic numbers*. But what do we mean by an algebraic number? They are the solutions to algebraic equations of the form:

$$A_0x^k + A_1x^{k-1} + A_2x^{k-2} + \cdots + A_k = 0,$$

where $A_0, A_1, \ldots, A_k$ are rational numbers, not all zero. However since an equation with rational coefficients can always be written as one with integral coefficients, we sometimes speak of the A's as integers. For example $\xi = \sqrt{2}$ is such an algebraic number because it is the solution of $x^2 - 2 = 0$. Of course algebraic numbers may be complex, but we shall restrict ourselves to the real ones.

We will show that the real algebraic numbers of the second degree are particularly obstinate — they refuse approximation of an order greater than 2. But what is the degree of an algebraic number? If ξ satisfies an algebraic equation of degree k but none of lower degree, then we say that ξ is an algebraic number of degree k. Here an example is helpful: $\sqrt{2}$ is of the second degree, although it satisfies the equation $x^2 - 4 = 0$. The polynominal $x^2 - 4$ can be factored as $(x^2 + 2)(x^2 - 2)$. As the latter can no longer be decomposed into factors with rational coefficients, we say that they are irreducible. Thus $x^2 - 2 = 0$ is the algebraic equation of the lowest degree satisfied by $\sqrt{2}$, so that $\sqrt{2}$ is of the second degree.

Let $\mathfrak{F}(x)$ be an irreducible polynominal of degree k in the rational field (when we speak of a polynominal in a field, we mean that its coefficients lie in that field). And let ξ be a root of $\mathfrak{F}(x)$, i.e., $\mathfrak{F}(\xi) = 0$, so that ξ is an algebraic number. Since $\mathfrak{F}(x)$ is irreducible, $\mathfrak{F}(x)$ and $\mathfrak{F}'(x)$ can have no common roots, so that $\mathfrak{F}'(\xi) \neq 0$.[2]

Now we propose to show that if ξ is an algebraic number of degree k, i.e., a root of a polynomial $\mathfrak{F}(x)$ of degree k, irreducible in the rational field, it is unreasonable to expect simple approximations which are very good. What we shall do is choose m/n fairly close to ξ and then estimate $|\xi - m/n|$ to show that this difference is greater than a certain function of n. Since $\mathfrak{F}(x)$ is a polynomial, $\mathfrak{F}'(x)$ is continuous, so that we certainly can choose m/n so close to ξ that we have

---

[2]This follows from the repeated application of the Euclidean algorithm. By successive long divisions we can always find the factors that $P_0(x)$ and $P_1(x)$ have in common, where $P_0(x)$ is a polynomial of higher degree than $P_1(x)$, for:

$$P_0(x) = P_1(x)Q_1(x) + P_2(x)$$
$$P_1(x) = P_2(x)Q_2(x) + P_3(x)$$
$$\vdots \qquad \vdots \qquad \vdots$$
$$P_{l-2}(x) = P_{l-1}(x)Q_{l-1}(x) + P_l(x)$$
$$P_{l-1}(x) = P_l(x)Q_l(x).$$

In this process the degree of the remainders, $P_0, P_1, P_2, \ldots, P_1$, decreases with each step, so that finally at a certain moment we must reach a polynominal of degree zero, i.e., $P_1 = $ const. Suppose $P_0(x)$ and $P_1(x)$ have a common root, say ξ. Then all the polynominals will have the root ξ; for at any state, two terms of the equation are certainly divisible by $x - \xi$, so that the third must also be divisible by $x - \xi$. And the division must come out even at the stage $P_{l-2}(x) = P_{l-1}(x)Q_{l-1}(x)$, i.e., the constant $P_l = 0$. Now rolling the process backwards, we see that all polynominals must have the factor $P_{l-1}(x)$ in common. Therefore the common factors of $P_0(x)$ and $P_1(x)$ can always be discovered by rational operations.

In particular if $\mathfrak{F}(x)$ and $\mathfrak{F}'(x)$ have a common root, $\mathfrak{F}(x)$ cannot be an irreducible polynomial.

$|\mathfrak{F}'(x)| < 2|\mathfrak{F}'(\xi)|$ in the whole environment of ξ which we consider.[3] Applying the mean value theorem, we have

$$\frac{\mathfrak{F}(\xi) - \mathfrak{F}(m/n)}{\xi - m/n} = \mathfrak{F}'(\vartheta),$$

where ϑ is some number lying between ξ and m/n. As $\mathfrak{F}(\xi) = 0$, introducing absolute values, we rewrite this as:

$$\left|\xi - \frac{m}{n}\right| = \left|\frac{-\mathfrak{F}(m/n)}{\mathfrak{F}'(\vartheta)}\right| > \left|\frac{\mathfrak{F}(m/n)}{2\mathfrak{F}'(\xi)}\right|$$

$$= \frac{1}{2|\mathfrak{F}'(\xi)|}\left|A_0\left(\frac{m}{n}\right)^k + A_1\left(\frac{m}{n}\right)^{k-1} + \cdots + A_k\right|$$

$$= \frac{1}{2|\mathfrak{F}'(\xi)|}\frac{|A_0 m^k + A_1 m^{k-1} n + A_2 m^{k-2} n^2 + \cdots + A_k n^k|}{n^k}$$

$$= \frac{M}{2|\mathfrak{F}'(\xi)|n^k} \geq \frac{1}{2|\mathfrak{F}'(\xi)|n^k}.$$

For M, being the absolute value of a polynomial in integers which is surely not zero, must be at least equal to 1. Hence we have:

$$\left|\xi - \frac{m}{n}\right| > \frac{1}{2|\mathfrak{F}'(\xi)|n^k}$$

or the difference between a real algebraic number of degree k and any rational approximation to it is greater than a constant times the reciprocal kth power of the denominator. Thus we find that algebraic numbers resist approximation to any order higher than their degree.

As a specific example let us consider an algebraic number of the second degree, perhaps the most familiar, $\sqrt{2}$. We have just shown that $|\sqrt{2} - m/n| > \mathscr{C}/n^2$ for a certain $\mathscr{C}$. On the other hand, we know that the inequality $|\sqrt{2} - m/n| < 1/n^2$ can be satisfied infinitely often. Hence this shows that this last approximation cannot be improved very much. However, the cubic algebraic numbers may permit an approximation of the order $1/n^3$.

In Liouville's argument, we assumed that $|\mathfrak{F}'(x)| < 2|\mathfrak{F}'(\xi)|$ held in some small environment of ξ, i.e., in $|x - \xi| < \delta$.

Is it possible that some of the numbers which we have not considered provide approximations of higher order? Those we considered all fell in the interval $-\delta \leq \xi - m/n \leq \delta$, so that we have excluded the rational numbers such that $m/n < \xi - \delta$, or $\xi + \delta < m/n$, or more briefly, such that $|\xi - m/n| > \delta$. If n is sufficiently large, clearly the inequality $|\xi - m/n| > \delta > 1/2|\mathfrak{F}'(\xi)|n^k$ will be satisfied. Thus at the worst we may be able to find finitely many choices of m/n which violate this inequality. But this is enough to prove the assertion that

---

[3]The number 2, in the inequality, may be replaced by $a > 1$.

in general we cannot approximate an irrational number to an order higher than the second, i.e., we cannot always find an infinite sequence of such approximations. We have found that, if ω is irrational, the inequality $|\omega - m/n| < 1/n^2$ is satisfied by infinitely many n, while if ξ is an algebraic number of degree k, the inequality $|\xi - m/n| \leq 1/a|\mathfrak{F}'(\xi)|n^k$, $a > 1$ can at most be valid only finitely often.

Let us complete the example with $\xi = \sqrt{2}$. Here $\mathfrak{F}(x) = x^2 - 2$, $\mathfrak{F}'(x) = 2x$, and $\mathfrak{F}'(\xi) = 2\sqrt{2}$. Now in the interior of the interval where $|\mathfrak{F}'(x)| < a|\mathfrak{F}'(\xi)|$ holds, we know that $|\xi - m/n| > 1/a|\mathfrak{F}'(\xi)|n^k$. Let us choose $a > 1$ so that $a|\mathfrak{F}'(\xi)| = a2\sqrt{2} = 3$. Then in the interval $0 < 2x < 3$, i.e., in $0 < x < \frac{3}{2}$, we know that $|\sqrt{2} - m/n| > 1/3n^2$. Outside of this interval this inequality can be violated only in a finite number of cases. Taking the first excluded number, $\frac{3}{2}$, we have:

$$\left| \sqrt{2} - \frac{3}{2} \right| > 0.085 \ldots > \frac{1}{3n^2} = \frac{1}{3 \cdot 2^2} = \frac{1}{12} = 0.08\overline{3}.$$

Similarly,

$$\left| \sqrt{2} - \frac{5}{3} \right| > 0.2 > \frac{1}{3 \cdot 3^2} = \frac{1}{27} = 0.037 \ldots .$$

Hence the finitely many exceptions to the inequality $|\sqrt{2} - m/n| > 1/3n^2$ are in fact non-existent. However, from the proof based on Dirichlet's pigeon-hole principle we know that we can satisfy $|\sqrt{2} - m/n| < 1/n^2$ infinitely often.

Can this last inequality be improved? In the next chapter we shall show that an algebraic number can be approximated infinitely often by rationals such that

$$\left| \xi - \frac{m}{n} \right| < \frac{1}{\sqrt{5}n^2},$$

and indeed that $1/\sqrt{5}$ is the best factor that we can introduce into this inequality.

The main result of this chapter is the inequality

$$\left| \xi - \frac{m}{n} \right| > \frac{1}{2|\mathfrak{F}'(\xi)|n^k},$$

where ξ is an algebraic number of degree k and $\mathfrak{F}'(\xi) \neq 0$, which holds with, at most, finitely many exceptions. This inequality has been strengthened by the Norwegian Thue (1909) and the German Siegel (1921). The former proved that

$$\left| \xi - \frac{m}{n} \right| > \frac{\mathscr{C}}{n^{(k/2)+1}},$$

while the latter showed that

$$\left| \xi - \frac{m}{n} \right| > \frac{\mathscr{C}}{n^{\sqrt{k}+1}},$$

both inequalities holding with finitely many exceptions. Thus Liouville's result is weaker than Thue's, while for $k > 4$, Thue's inequality is weaker than Siegel's.

Liouville used his discovery to make a very important decision. Are there any numbers that are non-algebraic? We know that the rational numbers (algebraic numbers of degree 1) are everywhere dense (i.e., between two given rationals we can always find another). However, these certainly do not exhaust all the real numbers, for there are also the quadratic algebraic numbers, e.g., $\sqrt{2}$ and all its multiples, that must fit in between the rationals. Next we consider the cubic numbers, the quartic numbers, ... ; and we find that the system of algebraic numbers becomes more and more dense. How do we know numbers that are non-algebraic? Can we define, can we exhibit such a number?

Liouville used the main inequality of this chapter to show that such transcendental numbers do exist. (In mathematics transcendental means "non-algebraic" — throw out the religious and philosophic connotations.) Consider the number

$$\vartheta = 1 - \frac{1}{2^{1!}} + \frac{1}{2^{2!}} - \frac{1}{2^{3!}} + - \cdots .$$

This cannot possibly be an algebraic number. If it were, it would possess a degree, say k; and it would not be possible to approximate it infinitely often to a degree higher than k, for we have just seen that this property characterizes the algebraic numbers. But we shall show that we can always approximate ϑ to a degree higher than any arbitrary number, so that no degree k could make the Liouville inequality valid.

If ϑ were algebraic, the inequality $|\vartheta - m/n| > C/n^k$ would hold with finitely many exceptions. But consider the partial sums of the series defining ϑ:

$$1 - \frac{1}{2^!} + \frac{1}{2^{2!}} - \frac{1}{2^{3!}} + - \cdots + \frac{(-1)}{2^{n!}} = p_n = \frac{N}{2^{n!}},$$

which are clearly rational numbers. The difference $\vartheta - p_n$ is the remainder term of an alternating series and hence less in absolute value than the first neglected term, i.e.,

$$|\vartheta - p_n| < \frac{1}{2^{(n+1)!}} = \frac{1}{(2^{n!})^{n+1}} = \frac{1}{2^{n!}(2^{n!})^n},$$

which holds for all n. Any inequality of the sort $|\vartheta - m/n| > C/n^k$, or specifically $|\vartheta - p_n| > C/(2^{n!})^k$, where k is fixed will be violated infinitely often because in addition to $|\vartheta - p_n| < 1/2^{n!}(2^{n!})^n$, we will have $1/2^{n!} < C$ and $n > k$ for all $n \geq n_0$, where n_0 is taken sufficiently large.

Hence the Liouville inequality is violated infinitely often, no matter how large k may be. Hence the assumption that ϑ is an algebraic number is untenable, for there is no algebraic equation of any degree that it could satisfy.

Thus ϑ must be a transcendental number, and we are indeed sure that such numbers exist, for we have exhibited one of the type known as Liouville's transcendental numbers. Not only do they exist, but G. Cantor proved that practically all real numbers are transcendental. The fact is that the algebraic numbers are denumerable, while the real numbers are non-denumerable. Therefore the excess of the set of real numbers over the set of algebraic numbers, i.e., the set of transcendental numbers, must be non-denumerable.

To close this chapter we give another application of Dirichlet's pigeon-hole principle — this time to show that there are an infinite number of integral solutions of the so-called *Pell's equation*:

$$x^2 - Dy^2 = 1,$$

where D is assumed to be a square free integer, as any square factor in D can be united with y^2.

What importance does this equation have? In Chapter 2 we found that the factorization of the numbers $a + b\sqrt{-5}$ was not unique. These numbers are but one subclass of the more general numbers of the form $a + b\sqrt{D}$ which are studied in algebraic number theory. Let us restrict ourselves to the case where $D > 0$, i.e., where $a + b\sqrt{D}$ is real. In the field composed of these numbers, we shall certainly call $x + y\sqrt{D}$ an integer if both x and y are themselves ordinary integers. (Other numbers may be counted as integers in this field — but this definition suffices here.) The conjugate of $a + b\sqrt{D}$ is $a - b\sqrt{D}$. And as in Chapter 2, we define the norm of a number to be the product of that number by its conjugate, i.e.,

$$N(a + b\sqrt{D}) = (a + b\sqrt{D})(a - b\sqrt{D}) = a^2 - Db^2.$$

Now if we should have a solution to Pell's equation, $x + y\sqrt{D}$, then it would be an integer in this field with a norm 1, i.e., $N(x + y\sqrt{D}) = x^2 - y^2 D = 1$. But this can be written in the form:

$$(x + y\sqrt{D})(x - y\sqrt{D}) = 1,$$

or

$$x - y\sqrt{D} = \frac{1}{x + y\sqrt{D}},$$

i.e., the reciprocal of $x + y\sqrt{D}$ is its conjugate. But correctly interpreted this is even more surprising. Here we have an integer whose reciprocal is also an integer. Such numbers we call *units*. In the field of real numbers ± 1 have this property; in the field of complex numbers, in addition to ± 1, we also are familiar with $\pm i$ as units. Now if we can show that the solutions of Pell's equation are infinite in number, then we will know that there are an infinite number of units in the field formed by the numbers $a + b\sqrt{D}$, in the field of real quadratic numbers. Herein lies the importance of Pell's equation to the theory of algebraic numbers.

As u and t are frequently used as arbitrary variables, it is fitting that such a unit should be denoted by u or by $u + t\sqrt{D}$. We could easily show, as before, that the norm of a product is the product of the norms. Hence the units have the property that if ξ is a number of the field,

$$N(\xi \cdot u) = N(\xi) \cdot N(u) = N(\xi),$$

i.e., the norm of the product of a number by a unit is equal to the norm of the number itself. Let us form such a product:

$$\xi \cdot u = \left(x_1 + y_1\sqrt{D}\right) \cdot \left(u + t\sqrt{D}\right)$$
$$= (x_1u + y_1tD) + (x_1t + y_1u)\sqrt{D}$$
$$= x_2 + y_2\sqrt{D}.$$

If $t \neq 0$, $x_2 + y_2\sqrt{D}$ is definitely different from $x_1 + y_1\sqrt{D}$, and we may say that the unit carries the first number over into the second. But from the foregoing we have:

$$N\left(x_2 + y_2\sqrt{D}\right) = N\left(x_1 + y_1\sqrt{D}\right).$$

Thus a solution of Pell's equation in integers yields not only infinitely many units, but also infinitely many numbers of the field, all of which have the same norm. (This follows from the fact that if u is a unit, then any power of u is also a unit.)

Dirichlet reverses the argument that we have just given. He begins with many numbers, all of which have the same norm, and from them he distills the units that carry these numbers over into themselves. He finds solutions to Pell's equation by finding units of the field $a + b\sqrt{D}$, but these are the main interest of the equation.

Following him, we must first restrict ourselves to norms that are not too large. Our first application of Dirichlet's pigeon-hole principle yielded the result: The inequality $|x\omega - y| < 1/N$, where ω is irrational and $y = [x\omega]$, is satisfied by some integer $1 \leq x \leq N$ for all N. Since D is square free, $\sqrt{D}$ is irrational. By choosing $\omega = \sqrt{D}$ and interchanging x and y, we may write this result as:

$$\left|y\sqrt{D} - x\right| < \frac{1}{N},$$

where

$$1 \leq y \leq N.$$

Not only do we have a solution to this inequality, but we can find as many different solutions as we need simply by taking N larger and larger. Let us restrict our attention to the numbers satisfying this inequality. Then we can reapply the inequality as a tool to estimate the magnitude of their norm, i.e., to estimate $|x^2 - y^2\sqrt{D}|$.

We have:

$$\left|x^2 - y^2D\right| = \left|y\sqrt{D} - x\right| \cdot \left|y\sqrt{D} + x\right| < \frac{1}{N}\left|y\sqrt{D} + x\right|.$$

Clearly, as $D > 0$ and only the values of squares are of interest, there is no loss of generality in assuming that x and y are also positive. What can we say about $y\sqrt{D} + x$? Adding to it the negative of its conjugate, we have:

$$(y\sqrt{D} + x) + (y\sqrt{D} - x) = 2y\sqrt{D} ,$$

or

$$y\sqrt{D} + x = 2y\sqrt{D} - (y\sqrt{D} - x).$$

Now we wish to take the biggest chance possible, to make an estimate for $|y\sqrt{D} + x|$ which will never be exceeded. To this end we note that y is at most N and $|y\sqrt{D} - x| < 1/N$, so surely $|y\sqrt{D} + x| \leq 2N\sqrt{D} + 1/N$. This gives us the result we are seeking:

$$|x^2 - y^2 D| < \frac{1}{N}\left(2N\sqrt{D} + \frac{1}{N}\right) = 2\sqrt{D} + \frac{1}{N^2}$$

or

$$|x^2 - y^2 D| < 2\sqrt{D} + 1 .$$

Thus if we solve $|y\sqrt{D} - x| < 1/N$, which we can always do by virtue of the pigeon-hole principle, we have consequently solved $|x^2 - y^2 D| < 2\sqrt{D} + 1$. But the former inequality has an infinite number of solutions (each of which yields a solution to the second inequality), $x + y\sqrt{D}$, whose norm, $x^2 - y^2 D = k$, is less in absolute value than $2\sqrt{D} + 1$, i.e., $-2\sqrt{D} - 1 < k < 2\sqrt{D} + 1$, where $k = 0$ is excluded, for if $x^2 - y^2 D = 0$, then D is not square free.

Thus we have found a way to restrict ourselves to norms that are not too large. Since there are only finitely many values that k may assume, while our process of finding these numbers gives us infinitely many of them, each of which has a definite norm, k, it follows that some value of k must reappear in the sequence, i.e., there will certainly be one k such that

$$x_a^2 - y_a^2 D = K = x_b^2 - y_b^2 D .$$

Moreover one value of k will be taken infinitely often. Specifically we classify the pairs of numbers x, y according to their norms k into one of the finitely many boxes:

$$-2\sqrt{D} - 1 < k < 2\sqrt{D} + 1 .$$

Since there are infinitely many pairs, at least one box, say K, will contain infinitely many pairs.

To this box we again apply the pigeon-hole principle. This time we compute the residues of x and y when divided by K and classify the pairs according to these residues. Thus we have K^2 possible boxes, one of which must again contain infinitely many pairs x, y which have the same remainders when divided by K.

Now we return to the reasoning with which we began this argument, and, taking any two solutions that have the property:

$$x_1^2 - y_1^2 D = K = x_2^2 - y_2^2 D ,$$

we try to run the reasoning backward and distill from these solutions the unit responsible for their relation, the unit that carries the first into the second. We have from our former reasoning two linear equations for u and t:

$$x_1 u + y_1 Dt = x_2$$
$$y_1 u + x_1 t = y_2,$$

which we can solve immediately.

$$u = \frac{\begin{vmatrix} x_2 & y_1 D \\ y_2 & x_1 \end{vmatrix}}{\begin{vmatrix} x_1 & y_1 D \\ y_1 & x_1 \end{vmatrix}} = \frac{x_1 x_2 - y_1 y_2 D}{K},$$

where the denominator is clearly $x_1^2 - y_1^2 D = K$; and

$$t = \frac{\begin{vmatrix} x_1 & x_2 \\ y_1 & y_2 \end{vmatrix}}{K} = \frac{x_1 y_2 - y_1 x_2}{K}.$$

But u and t should be integers. Dirichlet prepared this so carefully that these in fact are integers, as we shall easily see. Since both x_1 and x_2 leave the same remainder when divided by K, we have:

$$x_2 = x_1 + \mu K.$$

And similarly $y_2 = y_1 + \nu K$. Thus u turns out to be:

$$u = \frac{x_1(x_1 + \mu K) - y_1(y_1 + \nu K)D}{K}$$

$$= x_1 \mu - y_1 \nu D + \frac{x_1^2 - y_1^2 D}{K}$$

$$= x_1 \mu - y_1 \nu D + 1$$

an integer, while t is also an integer as

$$t = \frac{x_1(y_1 + \nu K) - (x_1 + \mu K)y_1}{K} = x_1 \nu - y_1 \mu.$$

So after all this preparation we have at hand an integral solution to Pell's equation. This can easily be verified. Not only do we have one, but we have actually found an infinite number of solutions. Moreover we have shown that there are an infinity of units and an infinity of numbers with the same norm in the field $a + b\sqrt{D}$. Herein lies the significance of the proof.

However, there is something disconcerting about this application of the pigeon-hole principle. Originally we had N boxes into which we distributed $N + 1$ objects, so that at least one box contained two objects. Here we have finitely many pigeon-holes into which we put infinitely many solutions. However, by examining the proof carefully, we can modify it so that we use the pigeon-hole principle in the first finite form and in such a way that we can in fact use the procedure described to construct one of the units whose existence was proved.

How many pigeon-holes did we use in the proof? In the first application of the pigeon-hole principle we had at most, being generous, the $2([2\sqrt{D}] + 1)$ boxes that corresponded to the values of k satisfying

$$-2\sqrt{D} - 1 < k < 2\sqrt{D} + 1$$

(recall $k = 0$ is excluded). Then in the second application, we had K^2 boxes corresponding to the possible residues when a pair, x, y, was divided by K.

Let us now from the very first subdivide each of the $2([2\sqrt{D}] + 1)$ boxes corresponding to the values of k into k^2 further pigeon-holes, i.e., so that we have the boxes listed below:

Value of k	Number of Boxes
± 1	1^2
± 2	2^2
± 3	3^2
$\vdots$	$\vdots$
$\pm([2\sqrt{D}] + 1)$	$([2\sqrt{D}] + 1)^2$

Since we have the same number of boxes for $-k$ as for $+k$, the total number is

$$2 \sum_{n=1}^{[2\sqrt{D}]+1} j^2.$$

But

$$\sum_{n=1}^{N} j^2 = \frac{N(N + 1)(2N + 1)}{6}.$$

Hence

$$2 \sum_{n=1}^{[2\sqrt{D}]+1} j^2 = \frac{1}{3}([2\sqrt{D}] + 1)([2\sqrt{D}] + 2)(2[2\sqrt{D}] + 1)$$

$$< \frac{1}{3}(2\sqrt{D} + 1)(2\sqrt{D} + 2)(4\sqrt{D} + 1)$$

$$< \frac{16D}{3}\left(1 + \frac{1}{2\sqrt{D}}\right)\left(1 + \frac{1}{\sqrt{D}}\right)\left(\sqrt{D} + \frac{1}{4}\right)$$

$$< \frac{16D}{3}(\sqrt{D} + 3).$$

This is certainly greater than the maximum number of boxes necessary to carry out the proof.

Now from the infinitely many pairs x, y which satisfy the inequality $|x^2 - y^2D| < 2\sqrt{D} + 1$, let us take just one more than $[(16D/3)(\sqrt{D} + 3)]$. These pairs we classify according to their norm and according to their residues when divided by their norm. Since we have at least one more pair than boxes, we certainly will have one box that contains at least two pairs. These have the property that they leave the same residues when divided by their norm, so that we can complete the proof as before. The special feature of this modified proof is that, given sufficient time, we could actually use it to construct a solution of Pell's equation.

If we have one solution of this equation, $u_1 + t_1\sqrt{D}$, then we actually have infinitely many. For if $N(u_1 + t\sqrt{D}) = 1$, then $N(u_1 + t_1\sqrt{D})^2 = 1$, so that

$$u_2 + t_2\sqrt{D} = (u_1 + t_1\sqrt{D})$$
$$= (u_1^2 + t_1^2 D) + (2u_1 t_1)\sqrt{D}$$

is also a solution. And as the result of a finer investigation, it turns out that there is one solution of Pell's equation that is the "smallest" and that all the others are powers of this "smallest."

Note to Chapter 7

Liouville's theorem has been improved successively by Thué (1909), Siegel (1921), Dyson (1947), and Gelfand (1952). Finally, Roth (1955) showed that if ξ is an irrational algebraic number and $\varepsilon > 0$, then there exists a constant $c_\varepsilon > 0$ (not effectively computable) such that for all relatively prime pairs of integers m, n,

$$\left| \xi - \frac{m}{n} \right| > c_\varepsilon\, n^{-2-\varepsilon}.$$

This very important result shows that no irrational algebraic number is approximable to any order greater than 2. It was asked by Serge Lang if the $n^{-\varepsilon}$ above could be improved to a negative power of log n.

Chapter 8

The Ford Circles

As one consequence of the pigeon-hole principle, we found that $|\omega - y/x| < 1/x^2$ can be satisfied by infinitely many x, no matter what irrational number may be chosen for ω. From this naturally arises the question: Can this inequality be improved? Liouville's theorem gives in part an answer and when applied to $\omega = \sqrt{2}$, yields a lower bound for α, $|\sqrt{2} - y/x| > \frac{1}{3}/x^2$ for all x. However, this leaves open the existence of an α between $\frac{1}{3}$ and 1 such that $|\omega - y/x| < \alpha/x^2$ may be satisfied infinitely often. If there is such an α, what is the best possible choice? In this chapter we shall find that $\alpha = \frac{1}{2}$ will do, and in fact that $\alpha = 1/\sqrt{5}$ is the best choice. Thus we must not only show that the inequality, $|\omega - y/x| < 1/\sqrt{5}x^2$ is satisfied infinitely often for all irrationals, ω, but also that there are some irrational numbers which resist a stronger inequality.

The fact that $\alpha = \frac{1}{2}$ will do has long been known and was discovered by representing the irrational numbers as continued fractions. However, such a proof requires too much preparation and technique to be given in these lectures. The discovery that $\alpha = 1/\sqrt{5}$ is the best value was made by A. Hurwitz in 1891. In this chapter we shall derive Hurwitz's result using an ingenious and elementary representation of fractions due to L. R. Ford (1938). (See Fig. 6.)

Suppose the fraction h/k, proper or improper, is reduced and $k > 0$. Let us lay out this fraction on a line, say the x-axis as in analytic geometry, and let us draw a circle with radius $1/2k^2$ which is tangent to the x-axis at the point h/k, i.e., a circle centered at the point whose coordinates are $(h/k, 1/2k^2)$. Such a circle of radius $1/2k^2$ tangent to the axis at h/k will be called the Ford circle corresponding to the fraction h/k and will be denoted by $\overline{h/k}$. The larger the denominator of a fraction, the smaller the radius of its Ford circle. (See Fig. 7.)

What is the distance between two Ford circles, $\overline{h/k}$ and $\overline{H/K}$, where $h/k \neq H/K$? As the center is the most representative point of a circle, it seems reasonable to interpret the "distance" between $\overline{h/k}$ and $\overline{H/K}$ as the distance, d, between their centers. Thus from elementary analytic geometry we have:

$$d^2 = \left(\frac{H}{K} - \frac{h}{k}\right)^2 + \left(\frac{1}{2K^2} - \frac{1}{2k^2}\right)^2.$$

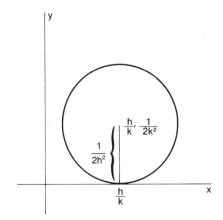

Figure 6

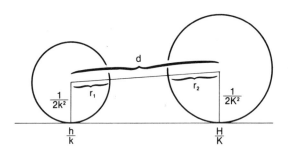

Figure 7

Let us compare d^2 with the square of the sum of the radii, i.e., form $d^2 - s^2$, where

$$s = r_1 + r_2 = \frac{1}{2k^2} + \frac{1}{2K^2},$$

$$d^2 - s^2 = \left(\frac{H}{K} - \frac{h}{k}\right)^2 + \left(\frac{1}{2K^2} - \frac{1}{2k^2}\right)^2 - \left(\frac{1}{2k^2} + \frac{1}{2K^2}\right)^2$$

$$= \left(\frac{H}{K} - \frac{h}{k}\right)^2 - \frac{1}{k^2K^2} = \frac{(Hk - hK)^2 - 1}{k^2K^2}.$$

Suppose $Hk - hK = 0$, then $H/K = h/k$ in contradiction to our assumption that H/K and h/k were different fractions. Hence $Hk - hK \neq 0$. And because all the numbers involved are integers, so that $Hk - hK$ is also one, we must have: $(Hk - Kh)^2 \geq 1$. So $(Hk - Kh)^2 - 1 \geq 0$, or $d^2 - s^r \geq 0$. This means that the distance between the centers of $\overline{h/k}$ and $\overline{H/K}$ is greater than or in the most unfavorable case equal to the sum of the radii. Thus *the Ford circles representing any two different reduced fractions cannot intersect — in the extreme case they may be tangent.*

These two circles will be tangent provided that $|Hk - Kh| = 1$. This is precisely the condition of adjacency for Farey sequences (Chapter 4). However, here it is useful to consider fractions as adjacent in a slightly broader sense. Now, we have infinitely many circles, none of which intersect — that is what we proved — which may at most be tangent to one another; and we call two fractions adjacent if their Ford circles are tangent, irrespective of the magnitude of the denominators. We shall find that any fraction has, in this sense, an infinitude of adjacents. (This was also the case in Chapter 4. If we had considered all the Farey sequences of order $n \geq k$, then we could have asserted that h/k had an infinity of adjacent fractions. However, we wish to make this chapter entirely independent of Chapter 4.)

How can we obtain all the fractions that are adjacent to h/k? There is no question that at least one adjacent exists as it can be shown in a variety of ways that the equation $xh - yk = 1$ always has solutions in integers. In fact this equation has an infinity of solutions, so that the problem is to find an appropriate expression for these adjacents. Suppose for the sake of definiteness that H/K is an adjacent to h/k and that another adjacent is denoted by H_n/K_n. What does this adjacency mean? Nothing more than the two equations:

$$hK - Hk = 1,$$
$$hK_n - H_n k = 1.$$

Subtracting, we have

$$h(K - K_n) - k(H - H_n) = 0,$$

or

$$h(K - K_n) = k(H - H_n);$$

and since from the outset we have been considering only reduced fractions, i.e., $(h, k) = 1$, we have:

$$k \quad \text{divides} \quad K - K_n$$
$$h \quad \text{divides} \quad H - H_n.$$

The first is easily written as: $K_n = K + nk$, where n is any integer, positive or negative, zero not excluded. Substituting this for K_n, we have:

$$-hnk = k(H - H_n)$$

or

$$H_n = H + nh.$$

Hence *all the adjacents to h/k may be expressed in the form $H_n/K_n = (H + nh)/ (K + nk)$, where $n = 0, \pm 1, \pm 2, \ldots$*. (Our seemingly curious notation for the adjacents, H_n/K_n, now has meaning, for n serves both as the integer used to compute them and as an index to distinguish them one from another.) Clearly these adjacents form two infinite sequences, one corresponding to $n < 0$, one to $n > 0$.

The particular disposition of fractions makes no difference because the essential point of the proof lies in the divisibility of $K - K_n$ and $H - H_n$ by k and h respectively.

Two adjacents to h/k whose indices differ by unity are themselves adjacent, i.e., H_n/K_n and H_{n+1}/K_{n+1} are adjacent. When we are concerned with reduced fractions, the equality of two fractions implies the identity of numerators and denominators. Thus if we knew that the expression for the adjacents to h/k which we just derived were in reduced form, we could use the preceding remark to show that

$$\begin{vmatrix} H_n & H_{n+1} \\ K_n & K_{n+1} \end{vmatrix} = \pm 1 .$$

Assuming this to be true, we can evaluate the determinant as follows:

$$\begin{vmatrix} H_n & H_{n+1} \\ K_n & K_{n+1} \end{vmatrix} = \begin{vmatrix} H + nh & H + (n+1)h \\ K + nk & K + (n+1)k \end{vmatrix} = \begin{vmatrix} H + nh & h \\ K + nk & k \end{vmatrix} = \begin{vmatrix} H & h \\ K & k \end{vmatrix} = \pm 1 .$$

But the last equality is merely a restatement of the hypothesis that H/K is adjacent to h/k. Moreover the argument shows that all the adjacents to h/k must be reduced, as no factor can be removed from any of the columns. Hence the Ford circles of successive adjacents are indeed tangent.

Now we have all the tools necessary to examine the geometry of this thing. In order to draw a definite picture (cf., Plate III), let us assume that the adjacent H/K is greater than h/k and $0 < K < k$. The circle $\overline{H/K}$ is tangent to and larger than $\overline{h/k}$. Among the adjacents to $\overline{h/k}$ there are two which are also adjacent to H/K, namely $H_1/K_1 = (H + h)/(K + k)$ and $H_{-1}/K_{-1} = (H - h)/(K - k)$ and the circles corresponding to these, $\overline{(H + h)/(K + k)}$ and $\overline{(H - h)/(K - k)}$, must be tangent to $\overline{h/k}$ and $\overline{H/K}$. These circles, $\overline{(H + h)/(K + k)}$ and $\overline{(H - h)/(K - k)}$, must be tangent to the x-axis and are thus completely determined by these three conditions of tangency. The circle corresponding to the fraction with the smaller denominator is the larger of the two, while the one which lies between $\overline{h/k}$ and $\overline{H/K}$ belongs to the fraction with the larger denominator. Continuing to the next pair of circles, $\overline{(H + 2h)/(K + 2k)}$ and $\overline{(H - 2h)/(K - 2k)}$, as they are adjacent both to the preceding circles and to $\overline{h/k}$, we see that they must in part fill in the chinks left between the x-axis and the circle $\overline{h/k}$. Stepwise we find that the adjacent circles form a complete ring about $\overline{h/k}$. The figure has been drawn for the case where $h/k < H/K$, $0 < K < k$, but it is clear that the argument would be the same if the inequalities were reversed.

From the foregoing it would seem that the two sequences of fractions adjacent to h/k actually converge to h/k as n becomes very large, positively or negatively, one from the left and the other from the right. To verify this, let us compute the difference between the abscissas of tangency of H_n/K_n and h/k.

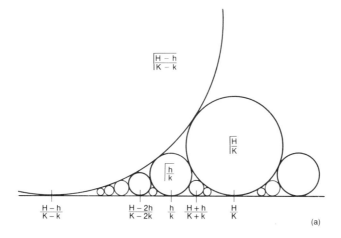

(a)

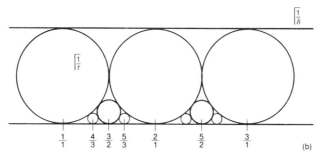

(b)

Plate III

$$\frac{H_n}{K_n} - \frac{h}{k} = \frac{k(H_n + nh) - h(K + nk)}{k(K + nk)}$$

$$= \frac{kH - hK}{k(K + nk)} = \frac{\pm 1}{k(K + nk)},$$

where the sign depends upon whether H/K is to the left or right of h/k. In the case we have considered in detail $h/k < H/K$ and the sign is plus: As n becomes very large positively, one sequence of adjacents approaches h/k from the right; while as n becomes large negatively, the other sequence of adjacents approaches h/k from the left. The fact that *the sequences of adjacents H_n/K_n approach h/k as n becomes large positively or negatively* is very important for the subsequent reasoning, for it shows that there is no gap in the chain of circles about h/k, except at the limit point, h/k, of the two sequences.

Now we can make our first conclusion. Let us draw a vertical line, $x = \gamma$, beginning high in the upper half plane, down to the x-axis. If γ should be rational, this line must indeed pass from some Ford circle directly to the axis at the point of tangency, $h/k = \gamma$. However, in the interesting case, when γ is irrational, the situation is different. Since the line $x = \gamma$ is irrational, it cannot pass directly to the axis from a Ford circle. It must leave every circle which it enters. However, every circle, $\overline{h/k}$, which it leaves is completely surrounded by a chain of adjacents; so it follows that the line $x = \gamma$ must enter another circle. This is true for all the Ford circles which it enters. Hence the line passes through an infinity of them. What does the passing of the irrational line, $x = \gamma$, through $\overline{h/k}$ mean? Clearly it means that the inequality

$$\left| \gamma - \frac{h}{k} \right| < \frac{1}{2k^2}$$

is satisfied. But the line passes through an infinite number of such circles, so that we have shown that if ω is irrational, the inequality $|\omega - h/k| < \alpha/k^2$ is satisfied infinitely often when $\alpha = \frac{1}{2}$. The Ford circles have so far enabled us to reduce the interval in which we wish to investigate α to $\frac{1}{3} < \alpha \le \frac{1}{2}$.

We have considered the geometry of these Ford circles only so far as was necessary to get the previous result. Now let us look into the matter more closely (cf., Plate III(b)). The simplest fractions are the integers, $n/1$. To these there corresponds a row of circles, $\overline{n/1}$, of radius $\frac{1}{2}$ tangent to the x-axis at the integral points. These are adjacent one to the next, and each has an infinitude of adjacents. However, as yet it is possible to escape from these circles in an upward direction without passing through an adjacent. To remedy this we introduce the line $y = 1$ as an improper circle, $\overline{1/0}$, which is tangent to all the circles, $\overline{n/1}$. This is reasonable, for a straight line may be thought of as a degenerate circle. Moreover $\left| \begin{smallmatrix} 1 & n \\ 0 & 1 \end{smallmatrix} \right| = 1$, so that this improper circle satisfies our adjacency condition. If $\overline{1/0}$ is a circle, degenerate though it may be, it should have an interior. This can be reasonably only the part of the upper half plane lying above $y = 1$, because as we go from the interior of a Ford circle through the point of tangency with an adjacent, we expect to pass into the interior of the adjacent.

Now for definiteness let us restrict ourselves to the interval between 1 and 2, as what occurs here will be repeated over and over in similar intervals. We expect to be able to draw a circle, $\overline{3/2}$, tangent to the x-axis at $\frac{3}{2}$ which is also tangent to both $\overline{1/1}$ and $\overline{2/1}$. Similarly we expect to be able to fit in between these three circles two others, $\overline{4/3}$ and $\overline{5/3}$, tangent to the x-axis at the points $\frac{4}{3}$ and $\frac{5}{3}$.

But do such circles actually exist? Can we construct them? This is the well known problem of Appolonius: Find a circle tangent to three given circles. Actually in our case there is quite a simplification as all three of the circles are tangent and one of them is degenerate.

The essential trick is to restate the problem by making a transformation known as inversion with respect to a circle (Chapter 9) which preserves both circles and angles, provided that straight lines are counted as degenerate circles and only the magnitude of the angle is in question. To make this transformation we introduce a circle of inversion of radius k (Fig. 8). Then a point at a distance r from the

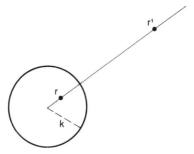

Figure 8

center of the circle is transformed into a point lying on the same ray from the center of the circle at a distance $r' = k^2/r$. Thus points lying on the circle of inversion go into themselves; and the interior of the circle is transformed into the exterior, and conversely. (The actual construction of these points is a matter of elementary geometry and is suggested by rewriting the definition of the inversion as $r'/k = k/r$.)

To simplify the problem let us choose a circle of inversion centered at a point of tangency and passing through a second point of tangency. This might be, for example, the dotted circle in Fig. 9. (In what follows descriptive terms such as "smaller," "larger," etc. are used only in reference to the specific figures discussed—they have no connotation of generality.) The straight line passing through the center of inversion will be clearly transformed into itself. The two points of intersection of the larger circle with the circle of inversion will go into themselves, and the origin which is also on this circle must go into the point at infinity, so that this larger circle becomes a second straight line parallel to the first. As both circles and angles are preserved, the third, smaller, circle will be transformed into a circle tangent to these two lines. Thus we have transformed the problem into the much simpler one of drawing a circle tangent to two straight lines and a circle touching both of them (Fig. 10). This has two easily constructed solutions. Reinverting these solutions, we find that we have two solutions to the problem, as we indeed expect (see Fig. 11).

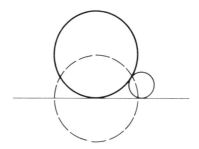

Figure 9

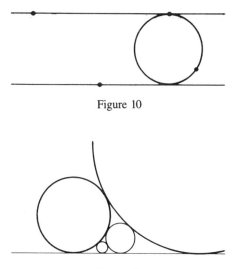

Figure 10

Figure 11

It may be of interest to sketch a solution to another easy case of the problem of Appolonius: Find a circle tangent to three circles, two of which intersect [cf., Plate IV(a)]. Our preceding luck suggests that we choose one of the points of intersection as the center of inversion and pass the circle of inversion through the other. This last point goes into itself, and the two intersecting circles, as they pass through the center of inversion, go into straight lines passing through this point which intersect at the same angles that the circle did. The third circle goes into another circle, say of radius r. Thus we are faced with the problem of drawing a circle tangent to two straight lines and a given circle [cf., Plate IV(b)]. Suppose the problem is solved. We see then that a circle with the same center and a radius increased by r would indeed pass through the center of the smallest circle and would be tangent to lines parallel to the given lines and backed off from the smaller circle by a distance r. If we draw these lines parallel and at a distance r from the two given lines, then we find that we can easily construct the solution to the revised problem of finding a circle tangent to two given lines and passing through a given point (the center of the circle of radius r). By reinversion then we have a solution to the original problem.

Thus we have given a geometrical, as well as an algebraic, argument for the existence of the Ford circles, and we are quite sure that our geometrical picture is correct. Cutting these circles by a vertical line $x = \omega$, ω irrational, yielded a profitable result.

What happens if we cut these circles by a horizontal line, $y = c \le 1$? Such a line cuts, or is tangent to, all circles whose diameter is $1/k^2 \ge c$, i.e., cuts all circles whose denominator $k \le 1/\sqrt{c}$. Here we have all fractions whose denominators are less than or equal to k arranged in their proper order, i.e., the fractions of the Farey sequence of order $k = [1/\sqrt{c}]$. From the foregoing argument it is

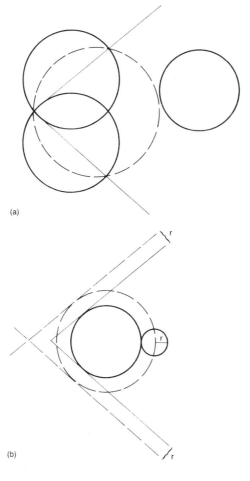

(a)

(b)

Plate IV

clear that all the Farey properties can be deduced from this geometrical representation and that Farey sequences can be constructed from the ordered fractions in any interval.

So far we have only examined a portion of our geometrical figure — we have directed our attention to the circles. These are all tangent and hence leave parts of the plane uncovered, namely the triangles formed by the arm of the Ford circles between their points of tangency. These uncovered areas we call *meshes*. If ω is irrational, the line $x = \omega$ which passes through an infinity of Ford circles must also pass through infinitely many meshes. Let us examine the reflection of such a line to a mesh (cf., Plate V).

The most obvious thing about a mesh is its corners. Let A be the point of tangency of the circles $\overline{h/k}$ and $\overline{H/K}$ which form a mesh together with a

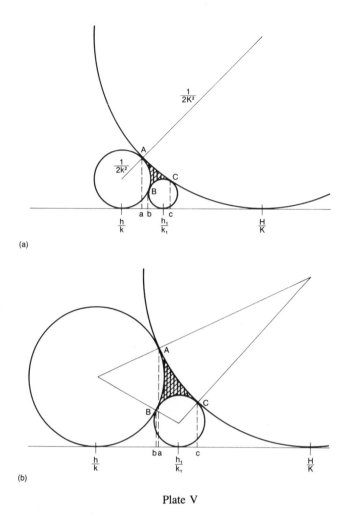

Plate V

third circle tangent to both of them. The point A divides the line connecting the centers of the circles $\overline{h/k}$ and $\overline{H/K}$ in the ratio $1/2k^2 : 1/2K^2=K^2 : k^2$. Hence the abscissa of A, a, lies at the point corresponding to the division of the line segment h/k, H/K in the same ratio, or:

$$a = \frac{k^2(h/k) + K^2(H/K)}{k^2 + K^2} = \frac{hk + HK}{k^2 + K^2}.$$

Thus the line $x = \omega$ cannot pass through the corner A, as the latter has a rational abcissa.

To be specific we have drawn the figure for the case $h/k < H/K$ and $0 < K < k$. As usual the argument will be entirely analogous no matter how the fractions are disposed in the inequalities.

To discuss the other corners of the mesh, we must know the third circle, call it $\overline{h_1/k_1}$. Since this circle is tangent to $\overline{h/k}$, it must be one of the

$\overline{(H + mh)/(K + mk)}$; since it is also tangent to $\overline{H/K}$, it must be one of the $\overline{(h + nH)/(k + nK)}$. This is only possible if $m = n = 1$, so that $h_1/k_1 = (h + H)/(k + K)$, the mediant of h/k and H/K. Now by a replacement of the letters in the preceding argument we easily write the abscissas of the other corners of the mesh, B and C, as:

$$b = \frac{hk + h_1k_1}{k^2 + k_1^2}, \qquad c = \frac{h_1k_1 + HK}{k_1^2 + K^2}.$$

Hence all the corners of the mesh are rational, so the line $x = \omega$ cannot pass through any one of them.

In deriving the inequality $|\omega - h/k| < 1/2k^2$ we used only the fact that the line $x = \omega$ passed through the circle $\overline{h/k}$. Now we know that this line must pass through a mesh — and more than that, through the interior of a mesh. This fact may help us find a better inequality. It is clear that the line $x = \omega$ will intersect the side of the mesh having the longest projection on the x-axis. Let the fraction whose circle forms this side of the mesh be denoted by $\mathcal{H}/\mathcal{K}$ and let $\mathcal{C}$ be the vertex of the mesh on this circle which is furthest away from the line $x = \mathcal{H}/\mathcal{K}$. If we knew which circle this were, we could find an inequality of the sort:

$$\left| \omega - \frac{\mathcal{H}}{\mathcal{K}} \right| < \mathcal{C} - \frac{\mathcal{H}}{\mathcal{K}}.$$

Can we do this? Can we estimate $\mathcal{C} - \mathcal{H}/\mathcal{K}$ so that it will be less than $1/2\mathcal{K}^2$ for all ω? We can, but to do so we must examine the relations between the vertices of the mesh in some detail.

First let us pay attention to the vertices lying on the larger circle. A must always be at, or below, the level of the center of this circle. C lies on the same arc as the circle and below A. Hence a and c cannot coincide, and $a < c$. Clearly $b < c$. Now what can we say about the relation of a and b? Let us form $b - a$. To simplify the computation we first find:

$$a - \frac{h}{k} = \frac{hk + HK}{k^2 + K^2} - \frac{h}{k} = \frac{kHK - hK^2}{k(k^2 + K^2)} = \frac{K}{k} \cdot \frac{kH - hK}{k^2 + K^2} = \frac{K}{k(k^2 + K^2)},$$

where the last step follows from the adjacency of the fractions. Similarly

$$b - \frac{h}{k} = \frac{k_1}{k(k^2 + k_1^2)}.$$

Subtracting these and recalling that $k_1 = k + K$, we have

$$b - a = \frac{k_1(k^2 + K^2) - K(k^2 + k_1^2)}{k(k^2 + K^2)(k^2 + k_1^2)}$$

$$= \frac{(k + K)(k^2 + K^2) - K(k^2 + \{k + K\}^2)}{k(k^2 + K^2)(k^2 + k_1^2)}$$

$$= \frac{k^3 - Kk^2 = kK^2}{k(k^2 + K^2)(k^2 + k_1^2)} = \frac{k^2 - kK - K^2}{(k^2 + K^2)(k^2 + k_1^2)}.$$

This does not look promising, but let us make a substitution in the numerator, namely, $h/k = s$. As we have assumed tht $0 < K \le k$, we always have $s \ge 1$. Clearly s is a rational number. Now

$$b - a = \frac{K^2(s^2 - s - 1)}{(k^2 + K^2)(k^2 + k_1^2)}.$$

This difference can never be zero. If it were, then $s^2 - s - 1 = 0$. Whence $s = (1 \pm \sqrt{5})/2$. But neither of these numbers is rational; whereas by our definition s is rational. Hence $b - a \ne 0$, i.e., b and a never can coincide.

On the other hand we can use the roots of this equation to characterize the different cases. We have:

$$b - a = \frac{K^2[s - (1 + \sqrt{5})/2][s + (\sqrt{5} - 1)/2]}{(k^2 + K^2)(k^2 + k_1^2)},$$

where $s + (\sqrt{5} - 1)/2 > 0$. Thus the sign of $b - a$ depends upon the sign of $s - (1 + \sqrt{5})/2$. Hence the two cases are characterized as follows:

$$a < b, \quad b - a > 0, \quad s - \frac{1 + \sqrt{5}}{2} > 0, \quad \text{or} \quad s > \frac{1 + \sqrt{5}}{2}, \quad \text{(I)}$$

$$a > b, \quad b - a < 0, \quad s - \frac{1 + \sqrt{5}}{2} < 0, \quad \text{or} \quad s < \frac{1 + \sqrt{5}}{2}. \quad \text{(II)}$$

In the first case when $a < b$, we are certain that ac is the longest projection of a side of the mesh, that the line $x = \omega$ intersects AC, and hence that $a < \omega < c$. This means that the line $x = \omega$ passes through the circle $\overline{H/K}$. However, we do not know through which of the two other circles it passes. Thus the appropriate comparison to make is:

$$\left| \omega - \frac{H}{K} \right| = \frac{H}{K} - \omega < \frac{H}{K} - a$$

$$= \frac{H}{K} - \frac{hk + HK}{k^2 + K^2} = \frac{Hk^2 - hkK}{K(k^2 + K^2)}$$

$$= \frac{k}{K} \cdot \frac{1}{k^2 + K^2},$$

as h/k and H/K are adjacent fractions. Now let us introduce $s = k/K$, so that we have:

$$\left| \omega - \frac{H}{K} \right| < \frac{s}{s^2 + 1} \cdot \frac{1}{K^2} = \frac{\xi(s)}{K^2}.$$

This should be better than our previous result. If so, the inequality

$$0 < \xi = \frac{s}{s^2 + 1} \le \frac{1}{2}$$

should hold. Transforming this, we have:

$$0 \leq s^2 - 2s + 1 = (s - 1)^2,$$

which is certainly true. We seem to be on the right track.

Our problem is now to investigate the behavior of $\xi = s/(s^2 + 1)$, for $s > (1 + \sqrt{5})/2 > 1$. Can we find an upper bound less than $\frac{1}{2}$ for this function in the interval we consider? Let us choose a second value of s, say $s' > s$, and form the difference:

$$\frac{s}{s^2 + 1} - \frac{s'}{s'^2 + 1} = \frac{ss'^2 + s - s's^2 - s'}{(s^2 + 1)(s'^2 + 1)} = \frac{(s - s')(1 - ss')}{(s^2 + 1)(s'^2 + 1)}.$$

Since $s \geq 1$, $1 - ss' < 0$. Hence the difference is positive. This tells us that the larger the value of the variable s, the smaller the value of the function ξ, i.e., ξ is a function that decreases steadily as s increases, a function that attains its maximum value for the smallest possible value of s. Hence we have in this case where $s > (\sqrt{5} + 1)/2$:

$$\xi = \frac{s}{s^2 + 1} < \frac{(\sqrt{5} + 1)/2}{\left[(\sqrt{5} + 1)/2\right]^2 + 1} = \frac{1}{\sqrt{5}}.\,^{1}$$

Thus for case (I) the inequality,

$$\left| \omega - \frac{H}{K} \right| < \frac{\xi}{K^2} \leq \frac{1}{\sqrt{5}K^2},$$

holds.

Now we come to case (II) [cf., Plate V(b)]. Here the geometry is slightly different in that the circles are more nearly of equal size. Here b lies to the left of a, so that BC is the side of the mesh with the longest projection on the x-axis, i.e., $b < \omega < c$. Hence the line $x = \omega$ passes through $\overline{h_1/k_1}$ and the appropriate comparison is $|\omega - h_1/k_1|$. Now we must choose between the two estimates of this difference, $|b - h_1/k_1|$ and $|c - h_1/k_1|$. The former is really the larger. Both B and C lie on the upper semicircle of $\overline{h_1/k_1}$. C must lie at least as high as B, if not higher, for we assume that $K \leq k$. Hence b lies further to the left of h_1/k_1 than c lies to the right of it. Therefore we are sure that the larger estimate is:

$$\left| \omega - \frac{h_1}{k_1} \right| < \left| b - \frac{h_1}{k_1} \right| = \frac{h_1}{k_1} - b = \frac{h_1}{k_1} - \frac{hk + h_1k_1}{k^2 + k_1^2} = \frac{h_1k^2 - hkk_1}{k_1(k^2 + k_1^2)}$$

$$= \frac{k(h_1k - hk_1)}{k_1(k^2 + k_1^2)} = \frac{k}{k_1(k^2 + k_1^2)},$$

as h_1/k_1 and h/k are adjacent fractions.

---

^{1}The foregoing follows a suggestion made by J. Massera.

As this case is characterized by $1 \le s < (\sqrt{5} + 1)/2$, we should like to introduce s into the right hand side of the estimate and reduce this to the form $\xi(s)/k_1^2$. Let us do it, recalling that $k_1 = k + K$.

$$\frac{k}{k_1(k^2 + k_1^2)} = \frac{kk_1}{k_1^2(k^2 + k_1^2)} = \frac{1}{k_1^2}\left(\frac{k(k + K)}{k^2 + (k + K)^2}\right)$$

$$= \frac{1}{k_1^2}\left(\frac{s(s + 1)}{s^2 + (s + 1)^2}\right) = \xi(s) \cdot \frac{1}{k_1^2}.$$

Thus we have $|\omega - h_1/k_1| < \xi/k_1^2$. And again the question is what can we say about ξ? Can we apply the same method as we used in case (I)? Let us choose a second value of s, say $s' > s$ and form the difference:

$$\frac{s(s + 1)}{s^2 + (s + 1)^2} - \frac{s'(s' + 1)}{s'^2 + (s' + 1)^2} = \frac{s(s + 1)}{2s(s + 1) + 1} - \frac{s'(s' + 1)}{2s'(s' + 1) + 1}$$

$$= \frac{s(s + 1) - s'(s' + 1)}{\{2s(s + 1) + 1\}\{2s'(s' + 1) + 1\}} = \frac{(s - s')(s + s' + 1)}{\{2s(s + 1) + 1\}\{2s'(s' + 1) + 1\}}.$$

Clearly $s + s' + 1 > 0$, so this difference is negative, i.e., to the small values of s belong the small values of ξ, and conversely. Hence the function ξ is increasing and its value in $1 \le s < (\sqrt{5} + 1)/2$ will be less than the value at $s = (\sqrt{5} + 1)/2$. Thus

$$\frac{s(s + 1)}{s^2 + (s + 1)^2} = \frac{s(s + 1)}{2s(s + 1) + 1} < \frac{[(\sqrt{5} + 1)/2][(\sqrt{5} + 3)/2]}{2[(\sqrt{5} + 1)/2]/[(\sqrt{5} + 3)/2] + 1} = \frac{1}{\sqrt{5}}$$

Hence

$$\left|\omega - \frac{h}{k}\right| < \frac{\xi}{k_1^2} < \frac{1}{\sqrt{5}k_1^2}.$$

In both cases (I) and (II) we have shown that if ω passes through the mesh in question, an approximation satisfying such an inequality does exist. But this is true for any of the infinitely many meshes through which the line $x = \omega$ passes. Hence we have proved Hurwitz's theorem: *For an irrational number ω, there exist infinitely many fractions h/k such that:*

$$\left|\omega - \frac{h}{k}\right| < \frac{1}{\sqrt{5}k^2}.$$

To show that Hurwitz's inequality is the best of its type, we must exhibit an irrational number for which the inequality $|\omega - h/k| < \eta/k^2$, where $0 < \eta < 1/\sqrt{5}$, can be satisfied by only finitely many h/k. The number, $\omega = (\sqrt{5} + 1)/2$, will not permit such an inequality to be satisfied infinitely often. Let us introduce this number into the inequality: $|(\sqrt{5} + 1)/2 - h/k| < \eta/k^2$, where $0 < \eta < 1/\sqrt{5}$, which we can rewrite, dropping the absolute value sign as:

$$\frac{\sqrt{5}}{2} + \frac{1}{2} - \frac{h}{k} = \frac{\vartheta}{\sqrt{5}k^2} = \pm\frac{\eta}{k^2},$$

where $|\vartheta|/\sqrt{5} = \eta < 1/\sqrt{5}$, or $|\vartheta| = \sqrt{5}\eta < 1$. The last inequality plays an important role in the proof. Putting all the $\sqrt{5}$'s on one side, we have:

$$\frac{1}{2} - \frac{h}{k} = \frac{\vartheta}{\sqrt{5}k^2} - \frac{\sqrt{5}}{2}$$

Squaring, we obtain:

$$\frac{1}{4} - \frac{h}{k} + \frac{h^2}{k^2} = \frac{\vartheta^2}{5k^4} - \frac{\vartheta}{k^2} + \frac{5}{4} - 1 - \frac{h}{k} + \frac{h^2}{k^2}$$

$$= \frac{\vartheta^2}{5k^4} - \frac{\vartheta}{k^2} - k^2 - hk + h^2 + \vartheta = \frac{\vartheta^2}{5k^4} > 0$$

Now the important fact is that $|\vartheta|$ is less than 1. Hence $-k^2 - hk + h^2 \geq 1$, for otherwise we should contradict ourselves, in that a square can never be less than zero, while this expression is surely an integer. Exchanging the left hand side of our equation and k^2, we can write it as

$$k^2 = \frac{\vartheta}{5(h^2 - hk - k^2 + \vartheta)}$$

If we replace the denominator by its smallest value, we can only increase the value of this fraction. Thus

$$k^2 \leq \frac{\vartheta^2}{5(1 + \vartheta)} \leq \frac{\vartheta^2}{5(1 - |\vartheta|)},$$

where the second step follows from $\vartheta \geq -|\vartheta|$. Furthermore $|\vartheta| = \eta\sqrt{5}$ so that we have:

$$k^2 \leq \frac{5\eta^2}{1 - \sqrt{5}\eta}$$

Thus if a fraction satisfies an inequality of the sort $|(\sqrt{5} + 1)/2 - h/k| < \eta/k^2$, where $0 < \eta < 1/\sqrt{5}$, its denominator must be less than a definite positive number, $5\eta^2/(1 - \sqrt{5}\eta)$. Therefore there can be only a finite number of solutions to this inequality.

Consequently we can assert that in general the inequality of Hurwitz,

$$\left| \omega - \frac{h}{k} \right| < \frac{1}{\sqrt{5}k^2},$$

is the best of all the inequalities which are satisfied infinitely often.

What is the special property of $(\sqrt{5} + 1)/2$ that makes it one of the most badly behaved of the irrationals? This can be most easily exhibited by the use of continued fractions. We can write:

$$\frac{\sqrt{5}+1}{2} = 1 + \frac{\sqrt{5}-1}{2} = 1 + \frac{1}{2/(\sqrt{5}-1)}$$

$$= 1 + \frac{1}{2(\sqrt{5}+1)/(\sqrt{5}-1)(\sqrt{5}+1)}$$

$$= 1 + \frac{1}{2(\sqrt{5}+1)/4} = 1 + \frac{1}{(\sqrt{5}+1)/2}.$$

But we can apply the same process to the $(\sqrt{5}+1)/2$ that appears in the denominator, so that

$$\frac{\sqrt{5}+1}{2} = 1 + \frac{1}{1 + [1/(\sqrt{5}+1)]/2}.$$

It is clear that we can continue this process indefinitely and develop this irrational number into an infinite continued fraction:

$$\frac{\sqrt{5}+1}{2} = 1 + \cfrac{1}{1 + \cfrac{1}{1 + \cfrac{1}{1 + \cfrac{1}{1} \cdots}}}.$$

In general if we have an infinite sequence of integers, $q_0, q_1, q_2, \cdots, q_n, \cdots$, we can arrange them in the scheme:

$$q_0 + \cfrac{1}{q_1 + \cfrac{1}{q_2 + \cfrac{1}{q_3 + \cfrac{1}{q_4 + \cfrac{1}{q_5} \cdots}}}}.$$

and form an infinite continued fraction. Here the q's are called the partial quotients of the continued fraction. The study of these fractions led to the first and still more usual solutions of the approximation problem that we have considered here and in the preceding chapter.

Let us look at the continued fraction for $(\sqrt{5}+1)/2$ more closely:

$$1 + \cfrac{1}{1 + \cfrac{1}{1 + \cfrac{1}{1 + \cfrac{1}{1} \cdots}}}.$$

If we cut it off short at a particular partial quotient, then we have an approximation to the irrational number that it represents. These rational numbers are known as the convergents of the fraction. Let us form these for our fraction:

$$1; \quad 1 + \frac{1}{1} = \frac{2}{1}; \quad 1 + \frac{1}{1 + 1/1} = \frac{3}{2};$$

$$1 + \frac{1}{1 + 1/[1 + (1/1)]} = \frac{5}{3};$$

and it is easy to see that the sequence continues $\frac{8}{5}, \frac{13}{8}, \frac{21}{13}, \cdots$. It can be shown that this sequence

$$\frac{1}{1}, \frac{2}{1}, \frac{3}{2}, \frac{5}{3}, \frac{8}{5}, \cdots \rightarrow \frac{\sqrt{5} + 1}{2}.$$

(The denominators of this sequence form the Fibonacci sequence, i.e.,

$$1, 1, 2, 3, 5, 8, 13, 21, \ldots$$

where each term is the sum of the two preceding terms.) Now the size of the partial quotients determines the rapidity with which the sequence of convergents approaches its limit. In the case of $(\sqrt{5} + 1)/2$ the quotients are as small as they can possibly be, so that the sequence of convergents could not converge more slowly.

A continued fraction all of whose partial quotients are positive, except possibly the first, is called regular (or simple). It turns out that a simple continued fraction whose sequence of partial quotients is periodic represents a quadratic irrational number. For example, suppose we did not know that the specific fraction we are considering represented such a number. Then from

$$w = 1 + \cfrac{1}{1 + \cfrac{1}{1 + \cfrac{1}{1 + 1 \cdots}}}$$

we have $w = 1 + 1/w$ (replace the second w by the continued fraction and we have the original continued fraction all over again). Thus $w^2 - w - 1 = 0$, or $w = (1 \pm \sqrt{5})/2$. But clearly the w represented by the fraction must be positive. So we have the result we secretly know, namely $w = (1 + \sqrt{5})/2$.

Problems

1. If u_n is the nth term of the Fibonacci sequence

$$1, 1, 2, 3, 5, 8, 13, 21, 34, \ldots$$

where $u_n = u_{n-1} + u_{n-2}$, show that $\lim_{n \to \infty} u_n/u_{n-1} = (\sqrt{5} + 1)/2$.

2. We can generalize the scheme of the Ford circles by keeping fixed the points of tangency of the circles with the x-axis and varying the radius of the circles. If we make the radius of these circles $1/\sqrt{3}k^2$ instead of $1/2k^2$, we will find that the meshes disappear, that the whole plane is covered by circles. A point will be covered at most by three circles, sometimes by two, and sometimes by only one.

   Show that if the radius is $1/\sqrt{3}k^2$, there will be no meshes, while if the radius is less than this, then meshes will exist. Where is the improper circle?

Chapter 9

On Linear Transformations —
The Modular Group

Mapping by inversion or by "reciprocal radii" forms the foundation for our work in this chapter. Inversion with respect to a circle of radius k is the transformation that carries a point at a distance r from the center of the circle into a point R on the same ray from the center such that $R = k^2/r$ (see Fig. 12). As we can return to the original figure merely by a similarity transformation (where all distances are magnified in the ratio $1/k$), there is no loss of generality in assuming that $k = 1$, i.e., we invert, now with respect to the unit circle, and our mapping is defined by $r \cdot R = 1$. Let us introduce a coordinate system with the origin at the center of inversion (the center of the circle) and let small letters denote the original points and large letters denote the image (transformed) points.

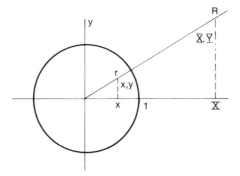

Figure 12

From the similarity of the triangles it is clear that $x/X = r/R$, or

$$X = \frac{R}{r}x = \frac{Rr}{r^2}x = \frac{x}{x^2 + y^2}$$

as $Rr = 1$ and $r^2 = x^2 + y^2$. Similarly $Y = y/(x^2 + y^2)$. However in this transformation there is no reason to prefer lowercase letters to capital ones — they play an equal role in the transformation — so that to each formula in small letters there corresponds one in capital letters:

$$x = \frac{X}{X^2 + Y^2}; \qquad y = \frac{Y}{X^2 + Y^2}$$

Thus, if we have a function of x and y which is equal to zero, say $f(x, y) = 0$, then the same function with x replaced by $X/(X^2 + Y^2)$, and y by $Y/(X^2 + Y^2)$ must also vanish, i.e., $f(X/(X^2 + Y^2), y/(X^2 + Y^2)) = 0$.

This remark can immediately be applied to show that this transformation preserves circles. The analytic expression of a circle is:

$$f(x, y) = A(x^2 + y^2) + Bx + Cy + D = 0.$$

From $r \cdot R = 1$ it follows that $x^2 + y^2 = 1/(X^2 + Y^2)$. Hence

$$f\left(\frac{X}{X^2 + Y^2}, \frac{X}{X^2 + Y^2}\right) = A\left(\frac{1}{X^2 + Y^2}\right) + \frac{BX}{X^2 + Y^2} + \frac{CY}{X^2 + Y^2} + D = 0.$$

Multiplying up, we have:

$$A + BX + CY + D(X^2 + Y^2) = 0$$

which is the equation of a circle in the transformed variables where the roles of A and D are interchanged. Two special cases should be mentioned:

1) If $A = 0$, the original circle is degenerate and we find that this straight line is transformed into a circle through the origin:

$$Bx + Cy + D = 0$$

becomes

$$BX + CY + D(X^2 + Y^2) = 0.$$

2) If $D = 0$, the original circle passes through the origin and its inversion is a straight line:

$$A(x^2 + y^2) + Bx + Cy = 0$$

becomes

$$A + BX + CY = 0.$$

These two cases may be visualized with the aid of Fig. 13, considering first the straight line and then the circle as the original figure to be inverted.

From the fact that the transformation is symmetrical in the small and large letters, it follows that the reinversion of a figure obtained by inversion yields the original figure again. The previous figure shows this very clearly. There the image of the straight line is the circle. Reinvert the circle and we have our straight line back again. A more geometrical reasoning is the following (see Fig. 14). Let us consider a circle and its image under inversion. From the definition of the transformation it is clear that the point of tangency, t, with a ray through the center

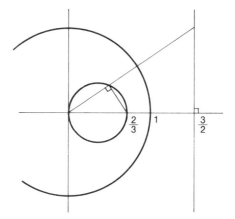

Figure 13

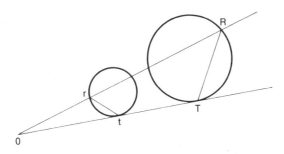

Figure 14

of inversion, 0, goes into another point of tangency, T. Any other point on the circle, say r, goes into a point R. Connecting r and t and R and T, we see that the two triangles so formed with their third vertex at 0 are similar. This follows from the fact that $rR = tT = 1$, or $r/t = T/R$. It is clear that the angle at r is equal to the angle at T. Constructions similar to this are frequently useful in the problems involving inversion.

We need now the following variant of the foregoing. Suppose the original circle passes through the center of inversion. Then we know that its image will be a straight line (see Fig. 15). Let d be the point diametrically opposite to 0 on the given circle, and D its image. Any point r on the circle goes into a point R on the straight line such that $r \cdot R = d \cdot D = 1$. Or $r/d = R/D$. Thus the two triangles are similar and in particular the angle at D must be a right angle, since we know the angle at r is right, and the angle at d must be equal to the angle at R.

This preliminary proposition enables us to show that inversion is conformal mapping: that is, it preserves angles in the small, i.e., locally, in the neighborhood of the vertex of an angle, the magnitude of the angle is preserved. Suppose our

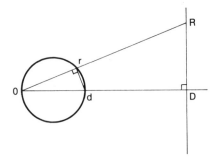

Figure 15

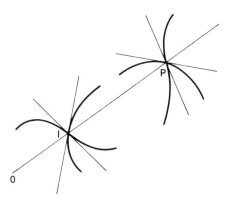

Figure 16

original figure consists of two intersecting curves (Fig. 16). Under inversion these will go into two other curves having as a common point P, the image of the original point of intersection, I. To describe the angle between the two curves we measure the angle between their tangents at the point of intersection. Thus the assertion that inversion is conformal is equivalent to saying that the angle between the tangents at I and at P is the same.

Now it is quite easy to prove this in the special case when the two original curves are circles; and from this we can easily deduce the general case. The two circles intersecting at I invert into two circles intersecting at P (Fig. 17). An original circle, b_1, can be replaced by another circle which has the same tangent at I and passes through the center of inversion. The latter circle inverts into a straight line which passes through the point P. What can this straight line be? Since it can only intersect the image of b_1 in the point P, it must be the tangent to the image circle at P. The angles 1 and 2, as well as the angles 3 and 4, fill out a right angle. But clearly angle 2 = angle 3. Hence angle 1 = angle 4. Changing our notation slightly, writing $I = r$ and $P = R$, we recognize in this figure our preliminary proposition. Thus angle 4 = angle 5 = angle 1.

However it is only fair to perform a similar replacement for the other original circle, b_2 (see Fig. 18). This leads us to the conclusion that angle b = angle d,

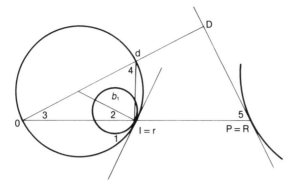

Figure 17

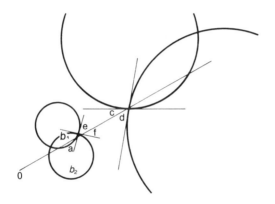

Figure 18

and we already know that angle a = angle c. Hence it follows that angle e + angle f = angle c + angle d. But this is what we set out to prove.

Since the angle between two curves is measured by the angle between the tangents at the point of intersection, we can replace the two curves of the general case by circles having the same tangents at the point of intersection. Therefore we can reduce the general case to the special case that we already know. Hence inversion is conformal.

Inversion is sometimes called *reflection* with respect to a circle. This is because of the fact that the associated points lie on the same ray from the center of the fundamental circle, i.e., the line connecting them is perpendicular to the circle of inversion. It is indeed clear that there must be a relation between reflection and inversion. And we can easily find out that reflection is the limiting case of inversion as the circle becomes very large while one point on it is kept fixed.

Moreover these two transformations have other properties in common. Clearly both inversion and reflection are transformations such that the reinversion or re-reflection yields the original figure again. Both transformations preserve the magnitude of the angles but change the sense of rotation (if we go from t_1 to t_2, where t stands for tangent, in the positive sense, after inversion for reflection we go from t_1 to t_2 in the opposite, negative, sense).

All the foregoing has been preparation for the discussion of linear transformations in the complex domain. In general we call a transformation $w = (az + b)/(cz + d)$ where w, z, a, b, c, d, are complex numbers, a linear transformation. The question that we ask in connection with such a function is: If z is permitted to run all around the complex plane, what can we say about the location of the point or number w? Clearly if $az + b = k(cz + d)$, $w = \text{const.}$ This is an uninteresting case as we have no association of points. To avoid it we forbid $\left|\begin{smallmatrix} a & b \\ c & d \end{smallmatrix}\right| = 0$, so we assume that $ad - bc \neq 0$.

We should keep it firmly in mind that we are in the complex plane, dealing with numbers of the sort $z = x + iy$, where x and y are bound together as parts of a single number. In analytic geometry we have used linear transformations of the sort:

$$u = \frac{a_1 x + b_1 y + c_1}{a_3 x + b_3 y + c_3},$$

$$v = \frac{a_2 x + b_2 y + c_2}{a_3 x + b_3 y + c_3},$$

where x and y are real numbers (this is a general projective mapping that sends straight lines into straight lines). Here we should distinguish the two real coordinates u and v from the single complex coordinates of a point, w, which contains the two real coordinates of a point bound together as a unit.

Let us proceed to examine the linear transformations in the complex plane systematically.

If $c = 0$, the transformation reduces to $w = (az + b)/d$, where $ad \neq 0$. This we can write as $w = (a/d)z + b/d = a_1 z + b_1$. Let us consider an auxiliary transformation, $Z = a_1 z$ where a_1 is a complex number. This is merely multiplication of one complex number, z, by another, a_1, which is a very simple matter. We merely have to enlarge everything by $|a_1|$ and then turn it around by the azimuth of a_1 (Fig. 19). Thus this auxiliary transformation consists of a dilation in the ratio $|a_1|:1$ and a turning through the angle equal to the azimuth of a_1. Thus the transformation $w = a_1 z + b_1$ consists of a dilation, a turning, and a shifting by vector b_1 — all of which are quite trivial operations.

Now let us consider the case $c \neq 0$. Let $\left|\begin{smallmatrix} a & b \\ c & d \end{smallmatrix}\right| = D$. Then we can write the linear substitution as

$$w - \frac{a}{c} = \frac{az + b}{cz + d} - \frac{a}{c} = \frac{bc - ad}{c(cz + d)} = \frac{-D}{c(cz + d)}.$$

This is practically the same as $W = 1/Z$ — it is only a shifting from the result of this.

$W = 1/Z$ consists of an inversion plus a reflection. We know how to divide in the complex plane: First we perform an inversion with respect to the unit circle,

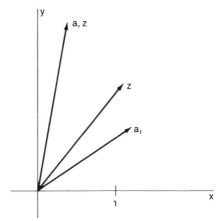

Figure 19

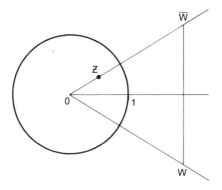

Figure 20

which carries the point Z into $\overline{W}$; and then we reflect with respect to the real axis, which carries the point $\overline{W}$ into the point W (see Fig. 20). (Multiplying Z by W we see that the result is 1, so that this operation is the inverse of multiplication.) Both inversion and reflection preserve circles and are conformal. Hence this transformation is conformal. Further the inversion changes the sense of rotation and the reflection changes it again, so that the combined operation clearly preserves not only the magnitude of the angle but the direction of rotation as well. Hence this transformation is fully conformal.

This is indeed the essence of the whole transformation, for to pass from W to w, we merely add a constant vector a/c. Now what is $Z = -c(cz + d)/D = -(c^2/D)(z + d/c)$? To obtain capital Z from z we merely have to make a shift of d/c and then introduce a constant factor $-c^2/D$, i.e., change the scale and increase the angle by a constant angle.

Thus the whole linear substitution is a very transparent thing: we know that every linear transformation of the complex plane can be composed from the four simple operations:

1) Shifting by a constant vector.
2) Changing the scale.
3) Rotation.
4) A combined operation of inversion-reflection — which is the operation of major interest.

This study of linear transformations will enable us to examine the configurations that will form the main subject of this chapter. Let us now consider the whole set of linear transformations of the form $w = (az + b)/(cz + d)$, where a, b, c, d are complex numbers such that $\left|\begin{smallmatrix} a & b \\ c & d \end{smallmatrix}\right| \neq 0$, and introduce the idea of a *group*. We can think of our transformations as being operations on the point z from which we derive another point w, which has the same nature as z. In order to have a group, first of all we must have a method of combining our operations such that the resulting operation also belongs to the group. Such a combination we shall call a *composition* of two operations.

How can we combine these linear substitutions? Clearly if

$$w = \frac{az + b}{cz + d}$$

and

$$W = \frac{a_1 w + b_1}{c_1 w + d_1}$$

are two linear transformations, we can form a third by performing these two successively, i.e.,

$$W = \frac{a_1[(az + b)/(cz + d)] + b_1}{c_1[(az + b)/(cz + d)] + d_1}$$
$$= \frac{(a_1 a + b_1 c)z + (a_1 b + b_1 d)}{(c_1 a + d_1 c)z + (c_1 b + d_1 d)} = \frac{Az + B}{Cz + D}.$$

Thus the two substitutions are combined into a third linear substitution whose coefficients are combinations of the old coefficients.

But how were the new coefficients, the capitals, obtained from the old ones, the small letters? If one has enough experience, this is recognizable as a case of the multiplication of matrices:

$$\begin{pmatrix} A & B \\ C & D \end{pmatrix} = \begin{pmatrix} a_1 & b_1 \\ c_1 & d_1 \end{pmatrix} \begin{pmatrix} a & b \\ c & d \end{pmatrix}$$
$$= \begin{pmatrix} a_1 a + b_1 c & a_1 b + b_1 d \\ c_1 a + d_1 c & c_1 b + d_1 d \end{pmatrix},$$

where the product is formed by multiplying rows by columns. (This type of multiplication first appears in the multiplication of determinants.) Now if we multiply two matrices, we have also multiplied the determinants of the matrices (don't confuse the two). Hence as we have assumed that the two matrices $\left(\begin{smallmatrix} a_1 & b_1 \\ c_1 & d_1 \end{smallmatrix}\right)$ and $\left(\begin{smallmatrix} a & b \\ c & d \end{smallmatrix}\right)$ are non-singular, i.e., that their determinants are not zero, it follows that the matrix $\left(\begin{smallmatrix} A & B \\ C & D \end{smallmatrix}\right)$ is also non-singular. Thus the composition of any two linear substitutions whose determinants do not vanish is not only formally another linear transformation, but it also is one in which we are especially interested, one whose determinant does not vanish.

But the fact that we remain in the realm in which we operate when we combine two transformations is not enough to insure a group. In addition to this we must have an associativity. Suppose we denote a linear substitution $w = (az + b)/(cz + d)$ by $L(z)$. Now if we have three such transformations $L_2(z)$, $L_1(z)$, and $L(z)$, from these we can form by composition the transformations $L_2(L_1(z))$ and $L_1(L(z))$ which we write shortly as $L_2L_1(z)$ and $L_1L(z)$. If we apply the first to $L(z)$ and apply $L_2(z)$ to the second we have:

$$L_2L_1(L(z)) \quad \text{and} \quad L_2(L_1L(z)) .$$

What we require is that these two transformations should be equal, i.e.,

$$L_2L_1(L) = L_2(L_1L) .$$

That is, provided we maintain the order in which the transformations are to be combined, it makes no difference which way we associate them, which way we group them in pairs [cf., Maxime Bôcher (1949)].

Thirdly a group must have a unit or identity element. Here $w = z$, a rather trivial transformation that leaves things as they were, has this role. Its matrix $\left|\begin{smallmatrix} 1 & 0 \\ 0 & 1 \end{smallmatrix}\right|$ is clearly non-singular.

Finally to each element in a group there must correspond a reciprocal element — that is, a transformation that puts things back as they were before the transformation to which it is reciprocal was applied. Clearly our transformations have this property as we can go from w to z as well as from z to w. All we have to do to find the reciprocal transformation is to solve $w = (az + b)/(cz + d)$ for z. This yields:

$$z = \frac{-dw + b}{cw - a} ,$$

whose matrix $\left(\begin{smallmatrix} -a & b \\ c & -d \end{smallmatrix}\right)$ has the same determinant as the original matrix and hence is non-singular. Thus the reciprocal element is a member of the group.

Now it is clear that the set of linear substitutions whose matrices are non-singular form a group in that they have the properties:

1) A composition of two linear substitutions is also a linear substitution — a member of the group.
2) The law of composition is associative.
3) There exists a unit element.
4) For each member of the group there exists a reciprocal element.

This whole thing becomes interesting when we single out sub-groups, and our goal is to discuss some of the properties of the modular sub-group. A sub-group consists of a smaller set of elements of the group, elements that have among themselves the group properties. Clearly they must have the same sort of closed-ness that the group itself has — that is, the element resulting from the composition of two members of the sub-group must be again a member of the sub-group. As the law of composition for the sub-group is the same as that for the whole group, the associativity is trivial. However the sub-group must contain the unit element, and to each member of the sub-group there must correspond another member of the sub-group which is its reciprocal.

There are very many interesting sub-groups of the group of linear trans-formations which are much studied in mathematics.

For example the transformations of the form $w = z + b$, where b is any constant, clearly form a sub-group — two shiftings of the plane are equivalent to a single shifting. This is called a congruence group in that it carries any figure into a congruent one (see Fig. 21).

From this sub-group we can single out a further sub-group by only permitting shiftings of $m \cdot b$ where b is a fixed number and $m = 0, \pm 1, \pm 2, \ldots$. Trans-formations of the form $w = z + m \cdot b$ take any strip of width b into similar strips of width b. This is an example of what we call a discrete group. By this we mean that the images of a point z cannot cluster around z — they must be separated by a finite distance, here by a distance of at least b. The former sub-group from which we extracted the present one is an example of a non-discrete group. There as b was arbitrary, $w = z + b$ carries the point z into image points that are as close to z as we like — we merely have to choose b small enough.

In the theory of elliptic functions we consider the group of transformations of the form $w = z + m_1\omega_1 + m_2\omega_2$, where ω_2/ω_1 is not real, i.e. ω_1 and ω_2 do not lie on the same straight line. These transformations carry a fundamental paral-

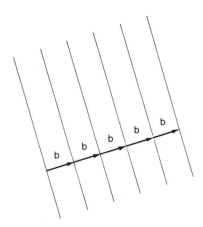

Figure 21

lelogram whose sides are the vectors ω_1 and ω_2 into similar parallelograms that cover the entire plane (Fig. 22). In the same manner the transformation $w = z + m \cdot b$ carries a single fundamental strip into similar strips that also cover the entire plane. Such a figure that can be shifted by a discrete group so that its images cover the entire plane is called the fundamental region of the group. In our examples the iteration of the process of the group covered the whole plane in the first case by a strip, in the second case by a parallelogram.

In the theory of complex variables these groups play an important role. For example we know the function $\sin z$ which has the property $\sin z = \sin(z + 2\pi)$. The transformation of z into $z + 2\pi$ is just such a shifting as we have been talking about in the discrete group $w = z + m \cdot b$. Thus we need to study this function only in the fundamental period strip. We could give many more examples.

But rather let us single out from our group sub-groups that have only a finite number of members. All of the groups that we have previously considered have had an infinite number of members. Clearly each finite group must contain the identity $w = z$, which has the non-singular matrix $\left(\begin{smallmatrix} 1 & 0 \\ 0 & 1 \end{smallmatrix}\right)$. If to this we add the operation $w = 1/z$, the inversion-reflection, we have a finite group. The matrix of this second member is $\left(\begin{smallmatrix} 0 & 1 \\ 1 & 0 \end{smallmatrix}\right)$ which is non-singular. Squaring this matrix, we have $\left(\begin{smallmatrix} 0 & 1 \\ 1 & 0 \end{smallmatrix}\right)\left(\begin{smallmatrix} 0 & 1 \\ 1 & 0 \end{smallmatrix}\right) = \left(\begin{smallmatrix} 1 & 0 \\ 0 & 1 \end{smallmatrix}\right)$, the matrix of the identity, so that we have confirmation of the fact we already knew: this transformation is its own reciprocal.

Another interesting sub-group is formed by multiplying z by the nth roots of unity, i.e., by the transformations $w = \rho^k z$, where $\rho = e^{2\pi i/n}$ (Fig. 23). Here we clearly do not need infinitely many powers of ρ, as these powers form the vertices of a regular polygon inscribed in the unit circle with n sides, and each transformation of the group corresponds to a rotation of the plane through an angle of $2\pi k/n$. Thus we only need the values $k = 0, 1, 2, \ldots, n - 1$; and we have a finite group consisting of n elements which has an infinite sector with central angle of $2\pi/n$ as its fundamental region.

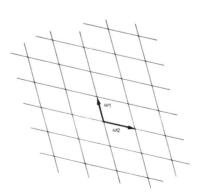

Figure 22

$$\rho = e^{2\pi i/6}$$

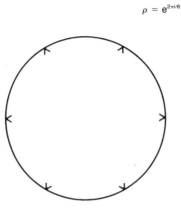

Figure 23

Now let us return to the linear substitutions with which we began. We have $w = (az + b)/(cz + d)$, where we assume that $\begin{vmatrix} a & b \\ c & d \end{vmatrix} = 0$. By introducing a constant λ such that

$$\begin{vmatrix} \lambda a & \lambda b \\ \lambda c & \lambda d \end{vmatrix} = \lambda^2 \begin{vmatrix} a & b \\ c & d \end{vmatrix} = 1,$$

we can write our transformation as

$$w = \frac{az + b}{cz + d} = \frac{\lambda az + \lambda b}{\lambda cz + \lambda d} = \frac{a_1 z + b_1}{c_1 z + d_1}.$$

Henceforward, we shall assume that $\begin{vmatrix} a & b \\ c & d \end{vmatrix} = 1$. We know that this mapping can be built up by a shifting, rotation, dilation, and an inversion-reflection.

So far we have neglected one source of information about the mapping — the derivative.

$$\frac{dw}{dz} = \frac{a(cz + d) - c(az + b)}{(cz + d)^2}$$

$$= \frac{ad - bc}{(cz + d)^2} = \frac{1}{(cz + d)^2}.$$

The absolute value of the derivative gives us the local enlargement, the local magnification at a point. Now for what points is this magnification 1? This will occur for the points which satisfy $|dw/dz| = 1/|cz + d|^2 = 1$ or $|cz + d| = 1$, provided $c \neq 0$, or $|z + d/c| = 1/|c|$, which is a circle centered at the point $-d/c$ of radius $1/|c|$ on the circumference of which the magnification is 1. This is called the *isometric circle*, i.e., the circle on which the measure is not changed.

Now what is the image of the isometric circle?

$$w - \frac{a}{c} = \frac{az + b}{cz + d} - \frac{a}{c} = \frac{bc - ad}{c(cz + d)} = \frac{-1}{c(cz + d)}.$$

Multiplying up by c, we have:

$$cw - a = \frac{-1}{cz + d}.$$

Hence

$$|cw - a| = \frac{1}{|cz + d|}.$$

Thus the isometric circle $|cz + d| = 1$, goes into the circle $|cw - a| = 1$, which has the same radius as: $|z + d/c| = 1/|c|$ and $|w - a/c| = 1/|c|$. This circle $|cw - a| = 1$ is the isometric circle of the inverse transformation $z = (-dw + b)/(cw - a)$. From the relation $|cw - a| = 1/|cz + d|$ it is clear that any point in the interior of $|cz + d| = 1$, i.e., those points for which $|cz + d| < 1$, goes over into the exterior of the circle $|cw - a| = 1$, and conversely a point in the interior of $|cw - a| = 1$ goes into the exterior of $|cz + d| = 1$.

This ends the generalities on linear transformations and now we begin a new line of thought concerning a special sub-group of the set of all possible linear transformations. So far we have merely imposed the condition that $\left| \begin{smallmatrix} a & b \\ c & d \end{smallmatrix} \right| = 1$, which was no restriction. Now let us impose the further restriction on our transformation that all the coefficients be integers, i.e., a, b, c, d shall henceforward be integers, positive, negative, or zero.

The first question we must answer is: Do these transformations form a "closed" set when we combine them? Suppose we have two such transformations,

$$w = \frac{az + b}{cz + d}, \qquad W = \frac{a_1 w + b_1}{c_1 w + d_1}$$

and we seek to write $W = (Az + B)/(Cz + D)$. Are the capital letters also integers? We found that we could write the matrix of the combined transformation as

$$\begin{pmatrix} A & B \\ C & D \end{pmatrix} = \begin{pmatrix} a_1 & b_1 \\ c_1 & d_1 \end{pmatrix} \begin{pmatrix} a & b \\ c & b \end{pmatrix}.$$

Clearly if the small letters are integers, so are the large ones. Thus this proposed sub-group satisfies the law of composition. The associativity is now trivial. Clearly the identity does belong to the group, for if $a = d = 1$, and $b = c = 0$, we have $w = z$; and the inverse also belongs as $z = (-dw + b)/(cw - a)$ is built out of integers. Thus the proposed sub-group fulfills all the requirements. This group of linear substitutions whose coefficients are integers is called the *modular group*.

The first thing we shall do with this group is to show that it has something to do with the Ford circles. We may note that this is an infinite group, for given any coprime a and b, we can always find an infinite number of solutions to $ax - by = 1$, each of which may serve as c and d, as we learned from the Farey sequences.

Now let $w = u + iv$, and $z = x + iy$, and let us seek the image of the line $v = 1$. From our general discussion of the linear transformation we know that the image of this line can only be another straight line or a circle. This was the main point of our discussion of inversion. Now we assume that $c \neq 0$, and we can clearly suppose $c > 0$. As we have done so often, we write our transformation as:

$$w - \frac{a}{c} = -\frac{1}{c^2(z + d/c)}.$$

Let $W = w - a/c$, $Z = c^2(z + d/c)$. Then this relation is $W = -1/Z$. If w runs along the line $v = 1$, the imaginary part does not change, so W runs along the same line at a distance $+1$ above the real axis. What we want to find is z, so first we find $Z = -1/W$. We know how division is carried out in the complex plane: First we must invert the line $v = 1$ with respect to the unit circle. It will go into a circle passing through the origin and tangent to the unit circle and also to the original line at $+1$. Now we must reflect this circle with respect to the real axis, which yields the dotted circle (Fig. 24). However we have forgotten the minus sign, which sends every point into one symmetrical with respect to the origin, so that finally the image of the line $v = 1$ turns out to be the circle in the upper half plane that passes through the origin and has a radius $\frac{1}{2}$. In Fig. 24 the successive positions of the point 1 under this series of transformations are marked by the numbers 1, 2, 3, and 4.

Now what can we say about $Z = c^2(z + d/c)$? Let $z^* = Z/c^2$. By this transformation the circle of radius $\frac{1}{2}$ becomes a circle of radius $1/2c^2$, touching the x-axis at the point zero. The final transformation $z = z^* - d/c$ is a shifting which moves the point of tangency of this small circle to $z = -d/c$. Thus we finally find that the line $v = 1$ is the image of a little circle in the z-plane, touching the x-axis at $-d/c$ and having a radius $1/2c^2$. This is an old friend — the Ford circle corresponding to the fraction $-d/c$. But as d/c can be taken as any rational number,

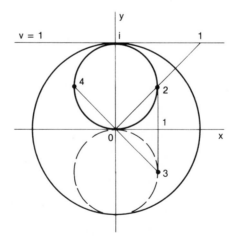

Figure 24

we see that we can obtain all Ford circles by transforming the line $v = 1$ by an appropriate substitution from the modular group. Since only c and d are specified by the fact that the transformation carries the line $v = 1$ into the Ford circle $\overline{-d/c}$, an infinite number of modular transformations will accomplish this, as there is an infinity of integer pairs a and b such that $\left|\begin{smallmatrix} a & b \\ c & d \end{smallmatrix}\right| = 1$, where c and d are given coprime integers.

Now what happens to the line $v = 1$ under the trivial transformations where $c = 0$, i.e., $w = (az + b)/d$. Here the determinant $\left|\begin{smallmatrix} a & b \\ o & d \end{smallmatrix}\right| = 1$, so that $a = d = 1$. Hence these transformations can be written as $w = z + b$. And they carry the line $v = 1$ into itself.

Thus we can assert that all the Ford circles including the improper one, $\overline{1/0}$, arise through applying transformations of the modular group to the line $v = 1$. Now we have seen that the modular group provides a background for the theory of Ford circles, and in fact the Ford circles have been developed from the consideration of the properties of the modular group.

What can we say about the image of an arbitrary point in the plane? First since the coefficients of the modular transformations are real, the real axis goes into itself, i.e., the image of a point on this line lies on the same line. A point in the upper half plane goes over into another point in the upper half plane. How can we see this? Let us write our transformation as:

$$w - \frac{a}{c} = \frac{-1}{c^2(z + d/c)} \quad \text{or} \quad W = -\frac{1}{Z}$$

What happens to a point, w, with a positive imaginary part, $\mathscr{I}(w) > 0$? Clearly if w lies in the upper half plane, so does W, for $\mathscr{I}(W) = \mathscr{I}(w) > 0$. But does Z also have a positive imaginary part? Let us perform the division. The inversion-reflection carries W into a point in the lower half plane, and then the minus sign carries this point into the upper half plane again (Fig. 25). The final transformation

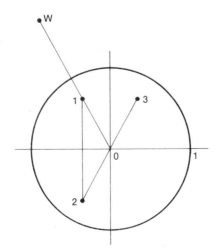

Figure 25

of Z into z is merely a shifting and a change of scale (as c is real, there is no rotation). Hence if w has a positive imaginary part, then z also lies in the upper half plane, and conversely. Now we could repeat the same argument for points in the lower half plane, and we would find that their images must also lie in the lower half plane. Thus under the modular transformations points in either half plane are kept separate from those in the other half plane — what happens in one half plane is repeated in the other — hence there will be no loss of generality if we forget the lower half plane and study these transformations only in the upper half plane. Sometimes we will include the boundary, the real axis, in the region which we consider, but more often we will exclude it and so eliminate difficulties arising from the special nature of the boundary points.

The modular group is discrete in the upper half plane, excluding the real axis. This means that about any image, w, of the point z, we can always draw a circle of finite radius that contains this one image, and no other images of z. We can prove this by applying a property of the isometric circles, namely: the interior of any isometric circle goes into the exterior of its image and conversely. Where do they lie? We can suppose that we have a full knowledge of the integers a, b, c, and d which are involved in the transformations we investigate. Each member of the modular group has an isometric circle of radius $1/|c|$ centered at the point $-d/c$, i.e., at a point on the real axis. If $c = 1$, the isometric circles, $|z + d| = 1$, are as large as possible with radius 1 centered at the integer points $-d$ on the real axis. All the isometric circles belonging to transformations having $c > 1$ will lie inside of these circles, which cover the upper half plane below the line $y = \frac{1}{2}\sqrt{3}$ completely (the shaded portion of Fig. 26). If $c = 2$, then d must be odd, for otherwise we could factor $\begin{vmatrix} a & b \\ c & d \end{vmatrix} = 1$. Thus the isometric circles belonging to $c = 2$ are of radius $\frac{1}{2}$ centered at the points $(2k + 1)/2$, $k = 0, \pm 1, \ldots, \pm n, \ldots$. Clearly the other isometric circles must be still smaller and they crowd down around the x-axis, growing smaller and smaller and more and more dense as c increases.

Now we are in a position to prove that the modular group is discrete. If the point z lies outside all the isometric circles, then its images lie either inside the isometric circles, or clearly at a finite distance one from another in the case where $c = 0$ so that $w = z + b$. In the former case it is also clear that the images of z cannot accumulate about any image, for there are only finitely many isometric circles that have a given point in common. On the other hand, if z lies inside the shaded region of the figure, it can lie in only finitely many isometric circles. For the remainder of the isometric circles the proof goes as before, and there are only

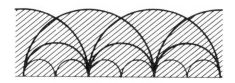

Figure 26

finitely many exceptional cases. However, in both cases, the points of the real axis are points of accumulation of the images of any point z. Thus the modular group is discrete in the upper half plane, provided that the real axis is excluded.

For the previous examples of groups we have found a fundamental region whose images under these transformations cover the entire plane. To express this differently we have found in each case a region into which any point in the plane can be transformed by a suitable member of the group. Now can we find such a region in the upper half plane for the modular group? As the interior of an isometric circle goes into the exterior of its image, which is also an isometric circle of the same radius, it is clear that such a region should lie outside all of the isometric circles. Now the transformations $w = z + b$ belong to the modular group, and in particular the transformation $w = z + 1$ is a member. Hence we see that the fundamental region must be of unit width and must lie outside all the isometric circles. Such a region can be formed by drawing vertical lines from the points of intersection of three successive isometric circles of radius 1 (the shaded region, $\mathcal{F}$, of Fig. 27). Now we must show that no point of $\mathcal{F}$ goes into another point of $\mathcal{F}$ under any modular transformation. (Actually some boundary points go into themselves by some modular transformations, but we shall overlook such exceptions for the moment.) The image of any point in $\mathcal{F}$ will go into a point in an isometric circle, if $c \neq 0$. However, if $c = 0$, then a point of $\mathcal{F}$ is subject to the transformation $w = z + b$, which carries z outside of $\mathcal{F}$, unless $b = \pm 1$ and z happens to be a boundary point. This difficulty we may eliminate by considering only a part of the boundary belonging to $\mathcal{F}$, or we may overlook it and deal only with the interior of $\mathcal{F}$.

Two points will be called *associated points* if they can be carried one into the other by some transformation of the modular group. Now we wish to show that any point in the upper half plane has one associated point in the fundamental region, i.e., that there exists a transformation of the modular group which carries this point into the fundamental region. If the point z_0 lies outside all the isometric circles, a transformation of the form $z' = z_0 + b$ will effect the desired movement, and there obviously can only be one such transformation.

Thus the only interesting case arises when the point z_0 lies in the interior of some isometric circles. Here it is advantageous to consider the part of the boundary of the largest isometric circle which is also common to $\mathcal{F}$ as a part of $\mathcal{F}$. We shall

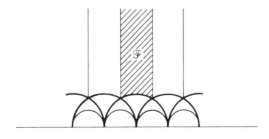

Figure 27

give two proofs of the fact there is a transformation of the modular group which does carry such a z_0 into the fundamental region. The first is an existence proof that hinges on the Bolzano–Weierstrass theorem and illustrates one of the most common "tricks" of proof used in mathematics. The second is a geometrical proof, one in which we could actually carry out the construction and find the image point.

Since z_0 is by assumption included in some isometric circle, we can draw a small circle about it of radius r_0 such that $r_0 < \delta$ where δ is the shortest distance from z_0 to the boundary of the isometric circle in which z_0 is included. What happens to this circle of radius r_0, which we assume to be entirely in the upper half plane, if we transform it by a member of the modular group in whose isometric circle it lies, i.e., by

$$Z' - \frac{a}{c} = \frac{-1}{c^2(z + d/c)},$$

where we can assume that $c > 0$. From our previous discussion we know it will go into a circle in the exterior of an isometric circle of the same size centered at $z' = a/c$ and that the image must lie in the upper half plane. Furthermore the image of the point z_0 will lie in the interior of this circle. But what can we say about the radius, r_1, of the image circle? To make a statement we must examine the transformation in greater detail. We know that $W = -1/Z$ is a combined inversion-reflection plus a symmetry with respect to the origin. If we carry this process through for one point, we see that it is equivalent to an inversion with respect to the isometric circle plus a reflection in a vertical line through $-d/c$. The remainder of the transformation

$$Z' - \frac{a}{c} = \frac{-1}{c^2(Z + d/c)}$$

is merely a shifting and a change of scale. Since reflection and shifting preserve the size of a figure, the inversion is the operation that will give us information about r_1.

Let us draw a tangent from $-d/c$ to these two circles intersecting them at distances R_0 and R_1 (Fig. 28). From the similarity of the triangles we have $r_0:r_1 = R_0:R_1$, or

$$r_1 = \frac{r_0 R_1}{R_0} = \frac{r_0 R_0 R_1}{R_0^2}.$$

But $R_0 R_1 = 1/c^2$, as they are inverse points. Furthermore $R_0 < 1/c - \delta$ and $\delta > r_0$, so that $R < 1/c - r_0$. Hence we have

$$r_1 > \frac{r_0(1/c^2)}{(1/c - r_0)^2} = \frac{r_0}{(1 - cr_0)^2} \geq \frac{r_0}{(1 - r_0)^2}.$$

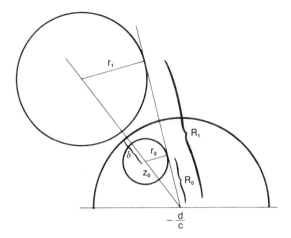

Figure 28

The factor $1/(1 - r_0)^2 > 1$, as $r_0 < \frac{1}{2}$, so that $r_1 > r_0$. There are now two possibilities:

1) This new circle lies outside all the isometric circles, so that the composition of $Z' = (aZ + b)/(cZ + d)$ and a transformation of the form $z'' = z' + b$ will carry z_0 into the fundamental region; and there is nothing more to prove.

2) This new circle lies inside some other isometric circle. Then there is another linear transformation belonging to this isometric circle that throws the circle of radius r_1 into the exterior of the image of this second isometric circle.

Under this transformation the radius of the image of our original small circle about z_0 becomes r_2; and repeating the previous argument, we have

$$r_2 > \frac{r_1}{(1 - r_1)^2} > \frac{r_0}{(1 - r_0)^2} \cdot \frac{1}{(1 - r_1)^2} > \frac{r_0}{(1 - r_0)^4}.$$

Again we have the same two possibilities. If we cannot now carry the image of z_0 into the fundamental region by $z'' = z' + b$, we repeat the process and find $r_3 > r_0/(1 - r_0)^6$. After k such steps we have $r_k > r_0/(1 - r_0)^{2k}$. Thus eventually we can make r_k larger than $\frac{1}{2}$ so that the image of the small circle about z_0 cannot lie, after some well defined step, entirely within the largest isometric circle.

If this process comes to a happy end, then we have the image of our small circle about z_0 lying entirely outside all the isometric circles. Then it needs only a final push, by shifting, to take it into the fundamental region. Since only a finite number of modular transformations are involved, by combining them successively, we have a definite modular transformation that carries z_0 into the fundamental region.

The more interesting case is that of the unhappy ending, where after a definite number of transformations the image of the small circle about z_0 straddles one of the largest isometric circles.

We only know that the image of z_0 surely lies in this circle; we do not know if it is in a position to go directly into the fundamental region. However, this is really enough to show that there is a transformation that carries z_0 into the fundamental region.

To do this we adopt the following device. We know that there are some points in the final image of the small circle about z_0 that are already in shape to go directly into the fundamental region. Take one of these, say z^*, which goes into z_1 in the small circle about z_0 and into z_1' in the fundamental region. Now draw a circle about z_0 that has z_1 in its exterior. With this new circle we can repeat the same process that we applied to the first circle. The process must come to an end as before and the new circle may either

1) lie entirely outside all the isometric circles so that there is nothing more to prove, or
2) straddle one of the largest isometric circles.

In the latter case we pick some point z_2 which is already in shape to go into $\mathscr{F}$ and has in $\mathscr{F}$ an image z_2'. Repeating this process again for a circle about z_0 that has z_2 in its exterior, we can find a point z_3 that goes into z_3' in $\mathscr{F}$. And if necessary we can define an infinite sequence of points $z_1, z_2, z_3, \ldots, z_n, \ldots$ which converge to z_0 and have images in $\mathscr{F}$, say $z_1', z_2', z_3', \ldots, z_n', \ldots$.

If we could show that this last infinite sequence of points lay in the finite part of the plans, then clearly the Bolzano–Weierstrass theorem would be applicable and we would be sure that they had at least one limit point. But this is clearly the case, as we assumed that the circle of radius r_0 about z_0 lay entirely in the upper half plane. Thus all the points $z_1, z_2, z_3, \ldots, z_n, \ldots$ and their images which still lie within the isometric circles are above a line $\mathscr{I}(z) = \varepsilon$, where ε is some small positive number. The image of this line will go furthest out in the plane when inverted with respect to the largest isometric circles. Clearly all z_n' must lie below the image of the line $\mathscr{I}(z) = \varepsilon$ when inverted with respect to the unit circle so that they are bounded. Thus the points $z_1', z_2', z_3', \ldots, z_n', \ldots$ converge to at least one limit point, say Z, which must correspond to z_0, under a certain transformation.

But what can we say about this transformation? We know that there is a modular transformation that carries z_k into z_k', say $z_k' = (a_k z_k + b_k)/c_k z_k + d_k)$. Is it possible that each z_k requires a new transformation? No. To see this we must look more closely at the process by which the z_k are defined. The point z_0, itself, and the small circles (neighborhoods) about it can lie in only a finite number of isometric circles, and the process we have described could have begun with any one of these and with any modular transformation which has the chosen circle as isometric circle. Further at each step of the process we could choose any modular transformation that has as isometric circle that circle in which the image of our neighborhood of z_0 lies. However once we have made these choices, we keep the transformations fixed. The case which interests us is that in which every neighborhood of z_0, no matter how small, finally straddles one of the largest isometric circles. Since the images of the neighborhoods of z_0 which we consider must pass through the image of the neighborhood of radius r_0 at the stage when it straddles the largest isometric circle, it is clear that only a finite number of isometric circles can be involved in the transformation of the points z_k into the points z_k'. The modular transformations which send z_k into z_k' arise from combining the modular transformations corresponding to these circles. But an isometric circle defines only the coefficients c and d of a modular transformation. Hence it is clear that only a finite number of c_k and d_k can enter into the transformations by which the sequence $z_1, z_2, z_3, \ldots, z_n', \ldots$ goes into the sequence $z_1', z_2', z_3', \ldots, z_n', \ldots$.

From this we can deduce that there are indeed only a finite number of transformations involved. Although from the specification of c and d, there may arise an infinite number of modular transformations, only one of these can be that one which carries our points into the fundamental region. To see this we ask what is the relation between two points z' and z'' arising from z by the modular transformations

$$z' = \frac{az + b}{cz + d} \quad \text{and} \quad z'' = \frac{a^*z + b^*}{cz + d},$$

which have the same c and d. Forming the difference we have

$$z' - z'' = \frac{(a - a^*)z + (b - b^*)}{cz + d}.$$

Since these are modular transformations,

$$\begin{vmatrix} a & b \\ c & d \end{vmatrix} = 1 \quad \text{and} \quad \begin{vmatrix} a^* & b^* \\ c & d \end{vmatrix} = 1$$

or

$$ad - bc = 1$$
$$a^*b - b^*c = 1;$$

hence $(a - a^*)d = (b - b^*)c$. But this means that $(a - a^*) = mc$ and $(b - b^*) = md$, where m is an integer. Thus $z' - z'' = m$, so that only one of the points arising from z by the transformations of the form $z = (az + b)/(cz + d)$, where c and d are fixed, can lie in the fundamental region. Since we know that all the z_k' lie in the fundamental region and that there are only finitely many c_k and d_k involved, it is now clear that there are only finitely many transformations involved.

Since by this finite set of transformations an infinite set of points $z_1, z_2, z_3, \ldots, z_n, \ldots \to z_0$ go over into another infinite set of points $z_1', z_2', \ldots, z_n', \ldots \to Z$, at least one transformation must be utilized infinitely often. We pick the infinite sub-sequence of points to which this transformation applies. This sub-sequence has the limit point z_0 and its images have the limit point Z. Hence as the linear transformations of the modular group are continuous, it follows that z_0 goes into Z by this transformation.

The sequence $z_1', z_2', \ldots, z_n', \ldots$ cannot have a second limit point Z_2 to which z_0 must also correspond, for then we should be able to define a transformation that carries the point Z_2 in the fundamental region into the point Z_1 which is also in the fundamental region.

Thus we have shown that there is a modular transformation that carries any point in the plane into a point of the fundamental region.

The second proof, that we give now, will introduce the generators of the modular group, which we must shortly discuss.[1] As any point lying in the exterior of all the isometric circles goes into the fundamental region by a transformation $z' = z + b$, we again are chiefly interested in the points which lie within some isometric circles. This time, however, we are content to look only at the largest isometric circles (Fig. 29). Any point that lies inside them must lie in a region similar to the shaded region, $A + B$ of the figure. If we can show that every point of this region has an image in the fundamental region, then we can apply the same process to any similar region.

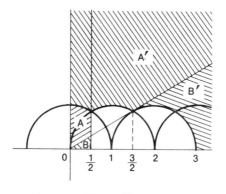

Figure 29

To do this we investigate the effect of the transformation $Z' = -1/Z$ on the points of this region. As we have pointed out before, this amounts to an inversion with respect to the unit circle plus a reflection with respect to the imaginary axis. As our goal is to move the point z_0 out of the isometric circles, it is clear that the inversion is the important operation. Consequently, we concentrate on it and forget the reflection. For simplicity let us assume that our point, z_0, lies in the isometric circle about zero, the unit circle, and that the real part of z_0 lies in $0 \leq x \leq \frac{1}{2}$. Let us draw from the origin the tangent to the isometric circle about 2. This tangent divides our region into two parts. Now let us see what happens to the upper part, A. The imaginary axis inverts into itself as it is a straight line through the center of inversion. Next the unit circle goes into itself. The straight line $x = \frac{1}{2}$ inverts into a circle passing through the origin and the points where $x = \frac{1}{2}$ intersects the unit circle. This must be the isometric circle about $z = 1$. Finally the tangent goes into itself. Since the interior of a figure goes into the interior of its image under inversion, we see that the image of a point in the upper part of our region, A, is already in shape to go into the fundamental region, as it lies in the region outside

[1]This follows a suggestion of J. Massera.

of all the isometric circles between the imaginary axis and the tangent, A' of the figure. Thus, an additional shift of the form $w = z + b$ will settle the whole affair.

The case in which the point z_0 lies below the tangent, i.e., in B, is a bit more difficult to settle. This region is bounded by the tangent, the line $x = \frac{1}{2}$, and the real axis. The real axis goes into itself. Hence the image of a point lying in this region must go into a point of the region between the real axis and the tangent which also lies outside the isometric circle about $z = 1$, i.e., into B'. Here we do not know if the image of z_0 lies above all the isometric circles or not. If it does not, we do know that its ordinate has been increased at least three times. The worst possible case would be the point which lies at the intersection of the tangent and the line $x = \frac{1}{2}$. This inverts into the point of tangency which has the abscissa $\frac{2}{3}$. Hence, by the similarity of the triangles its ordinate has also been increased thrice.

Any other point in this region will have its ordinate magnified in greater ratio. Now let us shift the point back into the strip between $-\frac{1}{2}$ and $+\frac{1}{2}$, by $z' = z + b$. Then we will have transformed the image of z_0 either into the fundamental region, or into a region symmetric to the one with which we began. In the latter case if it lies above the tangent, we can transform it by one more application of the preceding process into the fundamental region. If the image of z_0 still lies below the tangent, we know that it is at least three times further from the x-axis than it was originally. Hence by a finite number of applications of this process we can surely get an image of z_0 above the tangent, and thence transform z_0 into a point of the fundamental region.

Here we have shown that a given point z_0 can be transformed into the fundamental region by a finite combination of the modular transformations

$$z' = \frac{-1}{z} \quad \text{and} \quad z' = z + b.$$

Clearly the compositon of these will be another modular transformation.

The transformation $z' = z + b$ can be obtained from the transformation $z' = z + 1$ by combining the latter with itself b times in succession. So we see that the modular transformations can be generated from the composition of the two transformations

$$z' = \frac{-1}{z} = T(z),$$

$$z' = z + b = S(z).$$

By T^2 we shall mean the transformation T applied twice in succession:

$$T^2(z) = T(T(z)) = \frac{-1}{-1/z} = z.$$

Thus T^2 is actually the identity, so that we need only T. On the other hand T^0 also is the identity transformation, I. Similarly $S^0 = I$, but no other power of S equals I, for $S^q(z) = z + q$.

What we have shown is that any modular substitution, $M(z)$, can be written as a combination of powers of S and T, i.e.,

$$M(z) = S^{q_0}TS^{q_1}TS^{q_2}T \cdots TS^{q_k},$$

where $q_0, q_k \gtreqless 0$, and $q_1, q_2, \ldots, q_{k-1} \neq 0$. Suppose $q_2 = 0$. Then as $S^0 = I$, we could write:

$$
\begin{aligned}
M(z) &= S^{q_0}TS^{q_1}TTS^{q_3}T \cdots TS^{q_k} \\
&= S^{q_0}TS^{q_1}S^{q_3}T \cdots TS^{q_k} \\
&= S^{q_0}TS^{q_1+q_3}T \cdots TS^{q_k},
\end{aligned}
$$

so that if any of the interior q's are zero, we can combine the preceding and succeeding S's into one transformation. We now know that any point in the plane can be carried into the fundamental region by a modular substitution of the form $M(z)$. Before we give another proof of the fact that this is the most general modular transformation, let us investigate the fixed points of such transformations.

In general if two points z and z' are associated, there is a unique transformation that carries z into z', say $z' = M(z)$. But suppose that there is a second such transformation, say $z' = L(z)$, different from $M(z)$. These two transformations have inverses L^{-1} and M^{-1} which also belong to the modular group.

$$z' = M(z) = L(z).$$

Applying M^{-1}, we have:

$$z = M^{-1}M(z) = M^{-1}L(z),$$

as $M^{-1}M = I$, the identity. Let us write $M^{-1}L = N$, where N is a new modular substitution such that $z = N(z)$, i.e., the point z goes into itself under the transformation N. Such fixed points actually exist.

Now under what conditions can we have a fixed point? Suppose

$$\xi = \frac{a\xi + b}{c\xi + d},$$

where we can suppose $c > 0$ and $c \neq 0$. If $c = 0$, then the transformation reduces to $\xi = \xi + b$ and clearly b must equal zero, i.e., the only transformation of the form $w = z + b$ with fixed points is the identity and in this case every point is a fixed point. We can write this transformation as an equation for ξ, i.e.,

$$c\xi^2 - (a - d)\xi - b = 0,$$

which has the two roots

$$\frac{\xi_1}{\xi_2} = \frac{(a - d) \pm \sqrt{(a - d)^2 + 4bc}}{2c}.$$

These can be written, since $ad - bc = 1$, as

$$\frac{\xi_1}{\xi_2} = \frac{(a - d) \pm \sqrt{(a + d)^2 - 4}}{2c}.$$

Thus every modular transformation in which $c \neq 0$ has two fixed points, which may degenerate into a double point if $(a + d)^2 - 4 = 0$. Clearly if $a + d$ is of any size at all, the radical is positive. Now we must distinguish cases.

1) If $a + d = 0$, the fixed point in the upper half plane is

$$\xi = \frac{a - d + 2i}{2c} = \frac{-d + i}{c}$$

as $a = -d$. Let us take the most interesting case, $c = 1$, $d = 0$, so that $a = 0$. Then the matrix of this transformation is $\begin{pmatrix} 0 & -1 \\ 1 & 0 \end{pmatrix}$, which we recognize as that of T. Hence the transformation $T(z) = -1/z$ has the fixed point i.

This we can easily see geometrically since we know that $-1/z$ is a combination of inversion with respect to the unit circle and reflection with respect to the imaginary axis. The shaded half of the fundamental region (Fig. 30) goes into the shaded part of the "V" in the unit circle while the unshaded half goes into the unshaded part of the "V". And clearly the point i goes into itself. There are fixed points on every isometric circle which are images of the point i.

2) If $(a + d) = \pm 1$ (let us consider $a + d = +1$, as the argument for -1 goes in the same way), then the fixed point in the upper half plane is $\xi = (1 - 2d + i\sqrt{3})/2c$. Let us take $c = 1$, $d = 0$ again, so that $\xi = (1 + i\sqrt{3})/2$, which we recognize as one of the sixth roots of unity, which lies at the lower right corner of the fundamental region. It turns out that i and the two corners are the only fixed points in the fundamental region, for if we take c larger than one, then these fixed points will have no contact with the fundamental region.

The matrix of this transformation is

$$\begin{pmatrix} 1 & -1 \\ 1 & 0 \end{pmatrix},$$

i.e., $z' = (z - 1)/z = 1 - 1/z$, which we can write as $z' = ST(z)$. If we apply the transformation ST twice in succession, the resulting transformation has the matrix

$$\begin{pmatrix} 1 & -1 \\ 1 & 0 \end{pmatrix}\begin{pmatrix} 1 & -1 \\ 1 & 0 \end{pmatrix} = \begin{pmatrix} 0 & -1 \\ 1 & -1 \end{pmatrix}.$$

Applying ST again, we have

$$\begin{pmatrix} 0 & -1 \\ 1 & -1 \end{pmatrix}\begin{pmatrix} 0 & -1 \\ 1 & -1 \end{pmatrix} = \begin{pmatrix} -1 & 0 \\ 0 & -1 \end{pmatrix},$$

which for our purposes is equivalent to

$$\begin{pmatrix} 1 & 0 \\ 0 & 1 \end{pmatrix},$$

the matrix of the identity. Hence $(ST)^3 = I$. What does this mean geometrically? Let us follow the shaded half of the fundamental region, marked by 1 in Fig. 31. By T this goes into the region 2 and by ST, consequently, into the region 3. By inversion with respect to the unit circle and reflection in the imaginary axis 3 goes into 4 and by shifting into 5. This inverts and reflects into 6, which is shifted by S into the region 1 again. Thus it is clear geometrically that ST applied thrice is the identity.

This fact we can also see by combining the three transformations

$$z' = \frac{z - 1}{z}, \qquad z'' = \frac{z' - 1}{z'}, \qquad z''' = \frac{z'' - 1}{z'}$$

for we will find that $z' = z'''$.

The remaining two cases of the values of $a + d$ are very briefly dealt with.

3) If $a + d = \pm 2$, the fixed points coallesce and $\xi = (a - d)/2c$. This is not of much interest as ξ lies on the real axis which we have agreed to neglect.

4) If $|a + d| > 2$, then

$$\frac{\xi_1}{\xi_2} = \frac{a - d \pm \sqrt{(a + d)^2 - 4}}{2c}$$

are real quadratic numbers that also lie on the real axis.

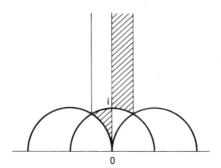

Figure 30

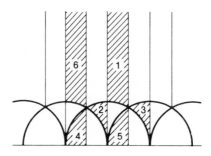

Figure 31

The most important part of this discussion of the fixed points was that we discovered the fact that $(ST)^3 = I$.

We give here a new proof of the fact that the most general modular substitution is of the form

$$M(z) = S^{q_0}TS^{q_1}T \cdots TS^{q_k},$$

where $q_1, q_k \gtreqqless 0$ and $q_1, q_2, \ldots, q_{k-1} \neq 0$. S is the transformation $z' = z + 1$ and T is the transformation $z' = -1/z$. If we have a modular substitution, we can write it in the following form:

$$\frac{az + b}{cz + d} = q_0 + \frac{(a - q_0 c)z + (b - q_0 d)}{cz + d} = q_0 + \frac{a_1 z + b_1}{cz + d}$$

$$= q_0 - \frac{1}{(-cz - d)/(a_1 z + b_1)}.$$

Clearly we can choose q_0 so that $|a_1| < |c|$, for a surely lies between two multiples of c. Now if we can do this for one modular transformation, we can do it for another, so that we have

$$\frac{-cz - d}{a_1 z + b_1} = q_1 + \frac{a_2 z + b_2}{a_1 z + b_1} = q_1 - \frac{1}{(-a_1 z - b_1)/a_2 z + b_2},$$

where $|a_2| < |a_1|$. And we can continue this process. At any stage we will have before us a modular substitution; take the first step for example, as

$$\begin{vmatrix} a - q_0 c & b - q_0 d \\ c & d \end{vmatrix} = \begin{vmatrix} a & b \\ c & d \end{vmatrix} = 1,$$

or

$$\begin{vmatrix} -c & -d \\ a_1 & b_1 \end{vmatrix} = \begin{vmatrix} a_1 & b_1 \\ c & d \end{vmatrix} = 1,$$

etc. Hence $|c| > |a_1| > |a_2| > \cdots$ is a succession of integers (coefficients of modular transformations), so that the process must come to an end with $a_{k+1} = 0$. This step will have the following shape:

$$\frac{-a_{k-1}z - b_{k-1}}{a_k z + b_k} = q_k + \frac{b_{k+1}}{a_k z + b_k}.$$

From the determinant

$$\begin{vmatrix} 0 & b_{k+1} \\ a_k & b_k \end{vmatrix} = 1$$

we conclude that $a_{k-1} = -b_{k+1} = 1$, so that the final fraction is of the form $-1/(z + b_k)$. This we can write as TS^{b_k}.

Now let us put this all together, starting from the very first. We have

$$M(z) = \frac{az + b}{cz + d} = q_0 - \frac{1}{M_1(z)}.$$

To obtain M from M_1 we must apply T to M_1 and then q_0 times in succession apply S to TM_1, or

$$M = S^{q_0}TM_1.$$

Similarly

$$M_1 = S^{q_1}TM_2.$$

Hence

$$M = S^{q_0}TS^{q_1}TM_2,$$

etc. Thus

$$M = S^{q_0}TS^{q_1}T \cdots TS^{q_{k-1}}TS^{q_k}TS^{b_k},$$

where q_0 and b_k may be zero, but $q_1, q_2, \ldots, q_k$ are not zero. Thus we have proved directly that M is the most general modular transformation and that any modular transformation can be generated by a suitable composition of S and T.

This expression of the modular transformation is clearly equivalent to the finite continued fraction

$$\frac{az + b}{cz + d} = q_0 - \cfrac{1}{q_1 - \cfrac{1}{q_2 - \cfrac{1}{q_3 - \cdots \cfrac{-1}{z + bk}}}}$$

which we must be careful to write with all minus signs, except for q_0 and b_k, in order that we have the proper sign for $T = -1/z$.

Now we have shown that $S = z + 1$ and $T = -1/z$ are generators of the modular group. However, using these two generators, we need all powers of S in order to generate the modular substitutions, while we use only one power of T. The fact that both $T^2 = I$ and $(ST)^3 = I$ suggests that we can find other generators for this group.

Let us keep T and introduce $U = ST$, which has the matrix $\left|\begin{smallmatrix} 1 & -1 \\ 1 & 0 \end{smallmatrix}\right|$ as the other generator. Here we should recall that $ST \neq TS$; this is not multiplication, but rather the composition of operations, the composition of transformations. If we multiply $U = ST$ by T from the right, we have $UT = STT = S$, as $TT = T^2 = I$. Hence we can replace S by UT in the preceding argument. Thus we see that we can form the substitutions of the modular group from the simpler generators, T and U: where

$T^2 = I$, so that T is needed only in the first power; and
$U^3 = I$, so that U is needed only in the first and second powers.

In this notation

$$M = U^{b_0}TU^{b_1}T \cdots TU^{b_{k-1}}TU^{b_k},$$

where $1 \leq b_1, b_2, \ldots, b_{k-1} \leq 2$ and $0 \leq b_0, b_k \leq 2$, all b's are integers. However, don't think that these two ways are the only ways in which the modular substitutions can be written, for there are other fixed points which yield transformations whose second or third powers are also the identity. For example, the point $(-1 + \sqrt{3}i)/2$ yields $(TS)^3 = I$, as $TS = TUT$, so that $(TUT)^3 = I$.

This last result could have been foreseen if we had written $(TUT)^3$ out in full:

$$TUTTUTTUT = \text{TUUUT} = TT = I, \quad \text{as} \quad TT = I \quad \text{and} \quad UUU = I.$$

It is by no means evident what the shortest way of writing such a modular substitution is. In group theory an expression such as $TUUUT$ is called a *word* and this question is, in part, equivalent to asking when two "words" have the same meaning, which is in general very difficult.

Another difficult question in group theory is the determination of the nature of a group from the nature of its generators. Not only must we know the generators, but we must also know the relations between them. A group is determined through its generators and their relations. As one example of this we have the modular group which has two generators T and U such that $T^2 = I$ and $U^3 = I$. Let us briefly make a few other examples.

If S is a single unrestricted generator, then we will have all powers of S in our group and we can write every "word" of this group as S^k. An example of such an infinite group arises when S means $z' = z + 1$.

However, if we restrict the generator of the group slightly, then the group becomes finite. Let V be a single generator of a group such that $V^n = I$. Then the group is finite and its "words" consist of the symbols $I, V, V^2, \ldots, V^{n-1}$. An example of such a group is that generated by $z' = \rho^k z$, where ρ is an nth root of unity, i.e., $\rho = e^{2\pi i/n}$. This is a finite group of rotations.

As we have just seen, we can form an infinite group from two generators of finite degree (T and U of the modular group are of finite degree in that $T^2 = I$ and $U^3 = I$). However, we can give an example of a group with two generators of finite degree which is finite. Let $T^*(z)$ mean $z' = 1/Z$ so $T^{*2} = I$, and $S^*(z)$ mean $1 - z$ so $S^{*2} = I$. Forming all possible combinations of S^* and T^*, we have only six elements in our group, namely:

$$z, \quad 1 - z, \quad \frac{1}{z}, \quad \frac{1}{1 - z}, \quad \frac{z - 1}{z}, \quad \frac{z}{z - 1}.$$

Let us make this as much like the modular group as possible; let us keep T^* as a generator and introduce $U^* = S^*T^* = 1 - 1/z = (z - 1)/z$ as a new generator, which has the matrix $\left(\begin{smallmatrix} 1 & -1 \\ 0 & 0 \end{smallmatrix}\right)$ (that of U of the modular group — however, not all the other substitutions of this new group belong to the modular group). Now we have $T^{*2} = I$ and $U^{*3} = I$. If these were the only relations between the generators, the "words" of this new group would be identical with the "words" of the modular group. However, we have now the additional relation $(U^*T^*)^2 = I$, for $U^*T^* = S^*$. The fact that this relation must be satisfied reduces the group from one of infinite order to one of finite order.

These examples illustrate a current problem in group theory: What can we say about a group when we are given its generators and the relations between them?

To close this chapter, we describe the modular figure in its entirety in a more complete manner than we have done heretofore. Let us see what happens to the images of the fundamental region when they are transformed into the unit circle by the transformation T. Since the fundamental region has symmetry with respect to the imaginary axis, let us shade half of it and mark the behavior of this one half. By shifting we can cover the plane above the isometric circles by images of the fundamental region. In Plate VI the fundamental region has been numbered 1 and its images near the unit circle have also been numbered. All these images will again have images within the unit circle which arise by the transformation T. These new images have been given numbers in Plate VI corresponding to the region from which they have been transformed (cf., Plate VI).

Since T consists of an inversion with respect to the unit circle plus a reflection in the imaginary axis, the fundamental region itself, 1, is transformed into 1 in the interior of the unit circle. The image of the fundamental region, 2, is bounded by two straight lines, $x = -\frac{1}{2}$ and $x = -1$, and the isometric circle about $x = -1$. We invert this with respect to the unit circle: The line $x = -\frac{1}{2}$ clearly goes into the isometric circle through the origin, centered at $x = -1$; the circle centered at $x = -1$ goes into the straight line $x = -\frac{1}{2}$ (i.e., the isometric circle about $x = -1$ and the line $x = -\frac{1}{2}$ are images one of the other under inversion with respect to the unit circle); and the line $x = -1$ goes into the isometric circle passing through $x = -1$ and the origin. Hence the region 2 inverts into the region two circled. A reflection with respect to the x-axis carries this into the shaded part of the unit circle labeled 2. Clearly all the other numbered regions can be matched with their images in the interior of the unit circle by a similar process.

Now from the region 1 in the unit circle we can obtain, by shifting, other images of the fundamental region exterior to this circle, such as 6, 7, 8, and 9. When these are subject to the transformation T, they fall into the areas with the corresponding numbers in the interior of the unit circle. Now the pattern in the interior of the unit circle begins to be clear: the unit circle is divided into a sort of curvilinear checkerboard by the isometric circles in its interior and these are alternately the images of the shaded part and then of the unshaded part of the fundamental region. By a parallel shifting, what occurs inside the unit circle can be reproduced in any isometric circle of the same radius — and the pattern so produced has an image in the interior of the unit circle. Hence the whole plane is covered by a checkerboard of images of the fundamental region.

But the argument that we have just given is not restricted to the unit circle — it will hold, with a suitable modular transformation, for any isometric circle, no matter how small. Thus we see that inside any isometric circle a pattern similar to that in the unit circle is repeated — that in the interior of any isometric circle there is an image corresponding to any image of the fundamental region in the exterior of this circle. This we may express briefly by saying that in any neighborhood of a point on the real axis there is a full image of the upper half plane exterior to this neighborhood.

We recall that among the Ford circles there was one exceptional one, $\lceil 1/0$, which was tangent to the Ford circle of radius $\frac{1}{2}$ at the integer points, $\lceil n/1 \rceil$. Earlier

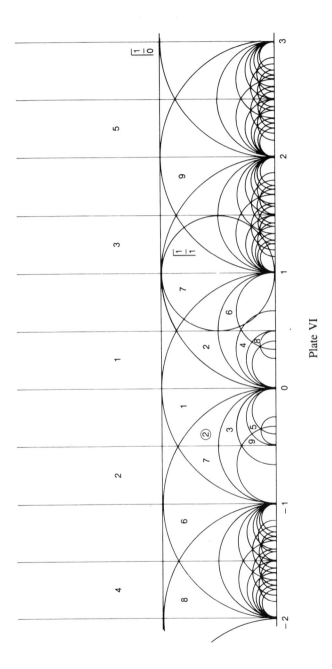

Plate VI

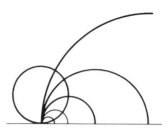

Figure 32

in this chapter we showed that all of the Ford circles could be derived from this one exceptional one by a suitable modular transformation. Let us draw this circle and its images, the proper Ford circles, in our figure. Now $\boxed{1/0}$ intersects the boundaries of the fundamental region at right angles, and since the modular substitutions are conformal mappings, all the images of the boundaries of the fundamental region, the isometric circles, will intersect the images of the proper Ford circles at right angles. This has been illustrated in Plate VI for $\boxed{1/1}$.

From our considerations of the isometric circles, we can see that through every rational point there pass infinitely many circles of the modular family (Fig. 32), but only one Ford circle. Now let us briefly consider the images of the rational points under a modular transformation $z' = (az + b)/(cz + d)$. Suppose $z = 0$; then $z' = b/d$. But by hypothesis b and d are coprime as the determinant $\begin{vmatrix} a & b \\ c & d \end{vmatrix} = 1$ cannot be factored. Hence given b and d coprime, we can find an infinity of pairs a and c such that $ad - bc = 1$. This means that the origin can be transformed into any rational point b/d by an infinite number of modular substitutions. Since the modular substitutions form a group, this implies that any rational number is the image of any other rational number. Thus it is clear that the behavior of the modular substitutions near the x-axis is indeed very complicated and that we avoid difficulty by pursuing it no further.

Chapter 10

Functions Belonging to Groups

This chapter will consist in the main of things we already know. For example the function $\sin z$ is periodic with a period of 2π, i.e.,

$$\sin z = \sin(z + 2\pi) = \sin(z + 2k\pi).$$

This we can express in the language we have just learned by writing $z' = z + 2k\pi$ so that $\sin z = \sin z'$. Clearly the sine function does not change when its argument is subject to a linear transformation belonging to the infinite group $z' = z + 2k\pi$. This is an example of a function belonging to the group $z' = z + k\omega$, where ω is the period. By a change of scale we can make $\omega = 1$, i.e., we consider $\sin 2\pi z = \sin 2\pi(z + k) = \sin 2z'$. Hence $\sin 2\pi z$ is invariant under the transformations of the group $z' = z + k$ (which reminds us of the transformation of the modular group which we called S^k).

A function is said to belong to a group of it is invariant under the transformations of that group. Thus $\sin 2\pi z$ belongs to the infinite group $z' = z + k$.

Can we find functions invariant with respect to finite groups? The simplest of these is the cyclic group, the group of finite rotations $z' = \rho^k z$, where $\rho = e^{2\pi i/n}$, $k = 0, 1, 2, \ldots, n - 1$. The function z^n belongs to this group. Let us verify this:

$$z'^n = (\rho^k z)^n = \rho^{kn} z^n = e^{2\pi i k n/n} z^n = z^n$$

as $e^{2k\pi i} = 1$.

In general if we have a certain set of linear transformations, $L_0, L_1, \ldots, L_n, \ldots$, under which a function $f(z)$ is invariant, these transformations form a group. Why is this so? Suppose we have two linear substitutions $L_1(z)$ and $L_2(z)$ under which a given function is invariant; then by hypothesis:

$$f(L_1(z)) = f(z) \quad \text{and} \quad f(L_2(z')) = f(z').$$

(To avoid complications we assume that the function and all the transformations involved have the same domain of definition, say the whole plane.) Now let $z' = L_1(z)$, so that we have $f(L_2(L_1(z))) = f(L_1(z)) = f(z)$. This clearly means that the composition of two linear transformations that leave the function invariant is again a transformation that leaves the function invariant. Since the invariant property of the transformations is not lost when they are combined, the set of all such operations has the first group property. The associativity and identity require-ments are clearly met as we are dealing with linear transformations. And the inverse operation must appear, for if $z' = L_1(z)$, then $L_1^{-1}(z') = z$ and we have $f(z) = f(L_1^{-1}(z'))$. Thus the set of all operations with respect to which a given function is invariant must form a group.

The idea of investigating the functions which are invariant with respect to groups has been exploited in the theory of automorphic functions. One of the goals of this chapter is to show that there exists a deeper background in group theory to some of the well known properties of the elementary functions.

If we know the group in advance, can we find automorphic functions that belong to this group? In case the group is finite, then this is indeed quite easy. Say we have n elements in our group, $L_0 = z$, the identity, $L_1, L_2, \ldots L_{n-1}$. All we need to do is to form a symmetric function of the transformations and we will have exhibited an automorphic function belonging to this group. For example let

$$f(z) = L_0(z) + L_1(z) + \cdots + L_{n-1}(z) .$$

If we assert that this is invariant with respect to the transformations of the group, this means that

$$f(z) = f(L_k(z)) = L_0(L_k(z)) + L_1(L_k(z)) + \cdots + L_{n-1}(L_k(z)) .$$

Now from the fact that the L's form a group, each of the combined operations L_iL_k belong to the group, i.e., $L_iL_k = L_j$. Multiplying on the right by L_k^{-1}, we have $L_i = L_jL_k$. Now k is fixed; the index i runs through the numbers 0 to $n - 1$, yielding a new transformation at each step; hence j runs through the same numbers although in a different order. That is, all of the L's appear in the second list in the form L_iL_k. Thus $f(z)$ is a function invariant under the given finite group. Of course we could have taken any other symmetric function of the L's, say

$$g(z) = L_0(z) \cdot L_1(z) \cdot \cdots \cdot L_{n-1}(z) .$$

Let us apply this to our favorite finite group, the cyclic group, where $z' = \rho^k z$, $k = 0, 1, 2, \ldots, n - 1$. Here we have

$$f(z) = z + \rho z + \rho^2 z + \cdots + \rho^{n-1}z = z(1 + \rho + \rho^2 + \rho^3 + \cdots + \rho^{n-1})$$

$$= z\frac{1 - \rho^n}{1 - \rho} = 0, \quad \text{as} \quad \rho^n = 1 .$$

$f(z) \equiv 0$ is a perfectly good automorphic function; in fact any constant is such a function for any group. However, we feel a bit cheated in that we got more than we wanted. But let us try

$$g(z) = z \cdot \rho z \cdot \rho^2 z \cdot \cdots \cdot \rho^{n-1}z = z^n \cdot p^{1+2+\cdots+n-1} = z^n \rho^{n(n-1)/2}.$$

Let us assume n to be odd; then $\rho^{n(n-1)/2} = e^{2\pi i[n(n-1)/2]} = 1$. Hence z^n is an automorphic function of this group which is not trivial.

The group that we are considering is known as a *cyclic group,* as all the elements of the group can be generated from a single element. In this case the elements of the group consist of the powers of the transformation $L_1(z) = \rho z$. The operation L_1 consists of multiplying by the complex number ρ. Thus, $L_2(z) = L_1L_1(z) = L_1(\rho z) = \rho^2 z$. Let us write two elements of this group so as to indicate this fact clearly:

$$L_j(z) = L_1^j(z), \qquad L_k(z) = L_1^k(z).$$

Combining these two, we have:

$$L_j(L_k(z)) = L_1^j(L_1^k(z)) = L_1^{j+k}(z).$$

Hence it follows that in this group, as distinguished from other groups, the order of composition is immaterial. Such groups are called *Abelian groups.* It turns out that every cyclic group has this property. We know another cyclic group, that is, one whose elements can all be generated by the repeated application of one of them, namely: the group of transformations $z' = z + b$, where b is an integer, which is generated by $z' = z + 1$. Here it is clear that the order in which the shiftings are performed is immaterial. Cyclic groups may indeed be infinite.

But we have had examples of groups that are non-Abelian. Recall the group of six elements:

$$L_0 = z, \qquad L_1 = 1 - z, \qquad L_2 = \frac{1}{z}, \qquad L_3 = \frac{1}{1 - z},$$

$$L_4 = \frac{z - 1}{z}, \qquad L_5 = \frac{z}{z - 1}.$$

Let us form the substitutions

$$L_2(L_5(z)) \quad \text{and} \quad L_5(L_2(z))$$

$$L_2\left(\frac{z}{z - 1}\right) = \frac{1}{z/(z - 1)} = \frac{z - 1}{z} = L_4 = L_5\left(\frac{1}{z}\right) = \frac{1/z}{1/z - 1} = \frac{1}{1 - z} = L_3.$$

Since the results are different, here the order of composition is material and this group is non-Abelian. (It turns out that this is the group of lowest order — with the smallest number of elements — that is non-Abelian.) Let us form the sum of these substitutions and see what kind of an automorphic function we get:

$$f(z) = z + 1 - z + \frac{1}{z} + \frac{1}{1 - z} + \frac{z - 1}{z} + \frac{z}{z - 1}$$

$$= 1 + \frac{z - 1 + 1}{z} + \frac{z - 1}{z - 1} = 3,$$

which is not very promising. Let's try

$$g(z) = z \cdot (1 - z) \cdot \frac{1}{1 - z} \cdot \frac{z - 1}{z} \cdot \frac{z}{z - 1} = 1.$$

Again we are successful, and again the result is not very revealing. Let us try a third time with $h(z)$, the sum of the squares of the transformations of the group.

$$h(z) = L_0^2 + L_1^2 + L_2^2 + L_3^2 + L_4^2 + L_5^2$$

$$= z^2 + (1 - z)^2 + \frac{1}{z^2} + \frac{1}{(1 - z)^2} + \frac{(z - 1)^2}{z^2} + \frac{z^2}{(z - 1)^2}.$$

This is a rational function, which clearly can assume arbitrarily large values — we have only to make the absolute value of z sufficiently large, or sufficiently small — so that it is not a constant. A little manipulation yields

$$h(z) = \frac{2z^6 - 6z^5 + 9z^4 - 8z^3 + 9z^2 - 6z + 2}{z^2(z - 1)^2}.$$

This function $h(z)$ will turn out to be invariant if we replace z by any member of the group, i.e., it is an automorphic function belonging to this group. From the symmetry of the coefficients we can see that this is the case if z is replaced by $1/z$.

For our first example, $\sin z$, which is an automorphic function of the group $z' = z + 2k\pi$, we found that we had a fundamental region, a strip of width 2π, such that any point in the plane had an associated point in this strip. Similarly we saw that an infinite sector was the fundamental region of the group formed by $z' = \rho^k z, \rho = e^{2\pi i/n}$. Now we can ask for the fundamental region of the group that we have just considered. Since z and $1/z$ both belong to the group, we see that we need only consider the interior of the unit circle. Since there are six transformations in the group, we expect six regions in the plane, three inside and three outside the unit circle. But the transformation $z' = 1 - z$ is included in the group. Hence, we expect a symmetry about the line $x = \frac{1}{2}$. Thus, the unit circle is to be reflected with respect to this line. The plane is divided by the line $x = \frac{1}{2}$ and the two circles into six parts, any one of which we may call the fundamental region (Fig. 33). However, we should expect a more symmetrical figure than this from such a simple group. This we can obtain by making a stereographic projection

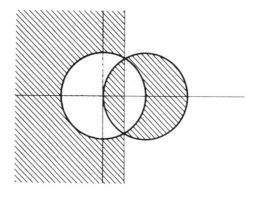

Figure 33

onto a sphere with the center of projection at the center of symmetry of the figure and with the sphere chosen of such size that the two points marked with heavy dots go into diametrically opposed points. This gives us an orange of six slices.

Now suppose we write $h(z) = f(z)/z^2(z - 1)^2$ and ask for the zeros of $h(z)$, or what is the same thing, the zeros of $f(z)$. If we know one, say ζ such that $f(\zeta) = 0$, then by the group property it follows that $L_1(\zeta)$, $L_2(\zeta)$, $L_3(\zeta)$, $L_4(\zeta)$, and $L_5(\zeta)$ are also roots of $f(z)$. Thus, we see that the roots of $f(z)$ all lie in the same field. Such an equation is called a *normal equation*.

Now having found automorphic functions belonging to the finite groups which we know, we ask for similar functions for the infinite groups that we know. For the infinite group $z' = z + 2k\pi$, we know that $\sin z$ is an automorphic function; but can we find others that are similar in structure to the ones which we found for finite groups? Can we find infinite combinations of the transformations of the group which are convergent and hence define automorphic functions for the group in question? Let $L_k(z) = z + 2k\pi$. Now $\sum_{k=-\infty}^{+\infty} L_k(z)$ is clearly not convergent. But let us try

$$\sum_{k=-\infty}^{\infty} \frac{1}{(z + 2k\pi)^2} = f(z) .$$

We know from the elementary theory of series that $\sum 1/k^2$ is convergent, and for large k,

$$\frac{1}{(z + 2k\pi)^2} \sim \frac{1}{4\pi^2 k^2} ,$$

so that $\sum_{-\infty}^{+\infty} (L_k(z))^{-2}$ will also be convergent. Now if we replace z by $z + 2\pi$, we have:

$$\sum_{-\infty}^{+\infty} \frac{1}{(z + 2\pi + 2k\pi)^2} = \sum_{-\infty}^{+\infty} \frac{1}{(z + 2(k + 1)\pi)^2} = f(z)$$

as both k and $k + 1$ run through all integers from minus infinity to plus infinity. Thus we have found by direct construction that $f(z) = f(z + 2\pi)$, i.e., $f(z)$, whatever it may be, is an automorphic function of the infinite group $z' = z + 2k\pi$. Those who have a little knowledge of function theory will recognize this as being $1/\sin^2(z/2)$.

Let us pass to the discussion of another group and show that such combinations of the transformations belonging to an infinite group can be made in more complicated ways. For example in the theory of elliptic functions we are concerned with those functions which are automorphic with respect to the transformations of the form $z' = z + m_1\omega_1 + m_2\omega_2$, where ω_1 and ω_2 are two complex numbers which are not colinear. The fundamental region in this case is a parallelogram with sides ω_1 and ω_2 whose images are the parallelograms with vertices at the points of the lattice generated by the vectors ω_1 and ω_2 (Fig. 34). Here we need more than the reciprocal second power of the members of the group. This time it turns out that $\sum_{-\infty}^{+\infty} (L_k(z))^{-3}$ is convergent. Let

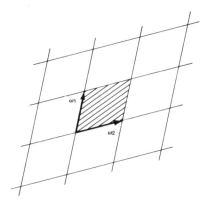

Figure 34

$$\Phi(z) = \sum_{m_1, m_2} \frac{1}{(z + m_1\omega_1 + m_2\omega_2)^3},$$

where the summation is to be taken over all values of m_1 and m_2, over all positive and negative integers including zero.

$$\Phi(2 + \omega_1) = \sum_{m_1, m_2} \frac{1}{(z + \omega_1 + m_1\omega_1 + m_2\omega_2)^3}$$

$$= \sum_{m_1, m_2} \frac{1}{(z + (m_1 + 1)\omega_1 + m_2\omega_2)^3}$$

$$= \Phi(z),$$

as m_1 and $m_1 + 1$ run over the same values, all the integers. In the same way we show that $\Phi(z + \omega_2) = \Phi(z)$. In $\Phi(z)$ we have an example of a function that has two periods in contrast with the elementary functions $\sin z$ and e^z which have a single period — 2π in the first case and $2\pi i$ in the second case.

Could either of the automorphic functions that we have just constructed turn out to be identically a constant? Let us look at the first example:

$$f(z) = \sum_{-\infty}^{+\infty} \frac{1}{(z + 2k\pi)^2} = \cdots + \frac{1}{(z - 2\pi)^2} + \frac{1}{z^2} + \frac{1}{(z + 2\pi)^2} + \cdots.$$

If $z \to 0$, the terms of the form $1/(z - 2\pi)^2$ will remain finite, but $1/z^2$ becomes as large as we please. Hence, $f(z)$ cannot be a constant. Similarly we can show that $\Phi(z)$ is not constant.

Since we have two examples of infinite groups whose elements can be combined into convergent combinations yielding automorphic functions, it is natural to ask if we can do the same with the modular group. H. Poincaré has given a recipe for forming a convergent sequence involving modular substitutions. How-

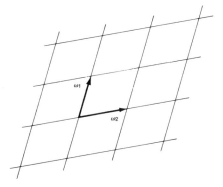

Figure 35

ever, we shall follow a recipe due to Eisenstein, a mathematician of the Nineteenth Century who died at twenty nine, most of whose work is incorporated into the Weierstrassian theory of elliptic functions.

Let us form the two series

$$G_2(\omega_1, \omega_2) = {\sum_{m_1, m_2}}' \frac{1}{(m_1\omega_1 + m_2\omega_2)^4},$$

$$G_3(\omega_1, \omega_2) = {\sum_{m_1, m_2}}' \frac{1}{(m_1\omega_1 + m_2\omega_2)^6},$$

where the prime indicates that the combination $m_1 = m_2 = 0$ is to be excluded from the summation, so that no denominator vanishes. These two series are absolutely convergent. (The sums of the odd reciprocal powers of $m_1\omega_1 + m_2\omega_2$ would also converge, but to the value zero as the negative terms would cancel the positive ones.)

Again we are dealing with the point lattice formed from the vectors ω_1 and ω_2, i.e., with the points $(m_1\omega_1 + m_2\omega_2)$ (Fig. 35). As before we agree that ω_1 and ω_2 should not be colinear and we agree to admit only such numbers ω_1 and ω_2 that we go from ω_2 to ω_1 in a counterclockwise direction. Thus, the argument (azimuth) of ω_1 is greater than the argument of ω_2, so that $0 < \arg \omega_1/\omega_2 < \pi$. Let $\omega_1/\omega_2 = \tau$. By our convention τ is a number in the upper half plane.

Let us consider our two series G_2 and G_3 at the same time by writing

$$G_k(\omega_1, \omega_2) = {\sum_{m_1, m_2}}' \frac{1}{(m_1\omega_1 + m_2\omega_2)^{2k}}.$$

These functions have the property that they are homogeneous in ω_1 and ω_2. If ω_1 and ω_2 are a pair of admissible values for these numbers, then so are $\lambda\omega_1$ and $\lambda\omega_2$ as $\tau = \lambda\omega_1/\lambda\omega_2$. Thus, if we replace ω_1 and ω_2 by $\lambda\omega_1$ and $\lambda\omega_2$, we have:

$$G_k(\lambda\omega_1,\lambda\omega_2) = {\sum}' \frac{1}{(m_1\lambda\omega_1 + m_2\lambda\omega_2)^{2k}}$$

$$= \lambda^{-2k}{\sum}' \frac{1}{(m_1\omega_1 + m_2\omega_2)^{2k}} = \lambda^{-2k}G_k(\omega_1,\omega_2).$$

and in analogy to elementary algebra we may say that G_k is a form of degree $-2k$.

Since our series are absolutely convergent, we can rearrange terms as we please provided that every point of the point lattice generated by ω_1 and ω_2 appears once and only once in the new summation. Of course the point zero is to be excepted. Let us look at our point lattice again. It seems to be formed by two vectors. But which pair is the fundamental pair? Clearly the fundamental pair is not uniquely defined as Fig. 36 indicates. However, if we pick a new pair of fundamental vectors, they certainly must lie among the old ones, i.e.,

$$\left.\begin{array}{l}\omega_1' = a\omega_1 + b\omega_2\\ \omega_2' = c\omega_1 + d\omega_2\end{array}\right\},$$

where a, b, c, and d are integers. However, it is not enough to say that the end points of the new vectors, ω_1' and ω_2', lie at the points of our lattice; we must be sure that all points of the lattice can be generated by linear combinations of these new vectors, that none of them are skipped. In particular we must be able to get back the old pair of fundamental vectors. If we can do this, then we are sure that no points of the original lattice will be missed by the combinations of the new generators. Thus, we require that:

$$\omega_1 = A\omega_1' + B\omega_2',$$
$$\omega_2 = C\omega_1' + D\omega_2',$$

where A, B, C, and D are integers. Let us write this out in full, using our expressions for ω_1' and ω_2', so that

$$\omega_1 = A(a\omega_1 + b\omega_2) + B(c\omega_1 + d\omega_2),$$
$$\omega_2 = C(a\omega_1 + b\omega_2) + D(c\omega_1 + d\omega_2),$$

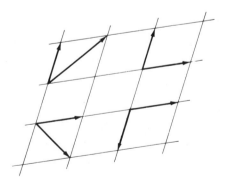

Figure 36

or

$$\omega_1 = (Aa + Bc)\omega_1 + (Ab + Bd)\omega_2\,,$$
$$\omega_2 = (Ca + Dc)\omega_1 + (Cb + Dd)\omega_2\,,$$

or finally

$$(Aa + Bc - 1)\omega_1 + (Ab + Bd)\omega_2 = 0\,,$$
$$(Ca + Dc)\omega_1 + (Cb + Dd - 1)\omega_2 = 0\,.$$

Since we know that ω_1 and ω_2 are not colinear, no linear combination of them can vanish unless the coefficients are identically zero. Thus, we have four linear equations which can be most easily expressed using matrix notation as

$$\begin{pmatrix} A & B \\ C & D \end{pmatrix}\begin{pmatrix} a & b \\ c & d \end{pmatrix} = \begin{pmatrix} 1 & 0 \\ 0 & 1 \end{pmatrix}.$$

Now the determinant of the matrix $\begin{pmatrix} 1 & 0 \\ 0 & 1 \end{pmatrix}$ is 1. Hence

$$\begin{vmatrix} A & B \\ C & D \end{vmatrix} \cdot \begin{vmatrix} a & b \\ c & d \end{vmatrix} = 1\,,$$

and this implies, as the elements of the determinants are integers, that $\begin{vmatrix} A & B \\ C & D \end{vmatrix} = \begin{vmatrix} a & b \\ c & d \end{vmatrix} = \pm 1$.

Let us look at the determinant $\begin{vmatrix} a & b \\ c & d \end{vmatrix}$, which is the determinant of the transformation of ω_1 and ω_2 into ω_1' and ω_2'. If we count the area of the parallelogram determined by ω_1 and ω_2 as positive and traverse this contour in the positive sense, then if $\begin{vmatrix} a & b \\ c & d \end{vmatrix}$ is positive, the area of the image parallelogram is also positive, i.e., we traverse it in the same sense as the original. Hence ω_1' and ω_2' are oriented in the same manner as ω_1 and ω_2. If $\begin{vmatrix} a & b \\ c & d \end{vmatrix}$ is negative, then we have misnamed ω_1' and ω_2'. Now if the names are changed, the rows of the determinant are interchanged and it becomes positive. Hence we assume that $\begin{vmatrix} a & b \\ c & d \end{vmatrix} = +1$, so that ω_1' and ω_2' are oriented in the same way as ω_1 and ω_2.

Now let us proceed to rearrange the terms of $G_k(\omega_1, \omega_2)$ according to the new fundamental vectors for our lattice, ω_1' and ω_2'. We know that $\begin{vmatrix} a & b \\ c & d \end{vmatrix} = +1$, which insures that no lattice points are missed.

$$G_k(\omega_1, \omega_2) = \sideset{}{'}\sum_{m_1, m_2} \frac{1}{(m_1\omega_1 + m_2\omega_2)^{2k}}$$

$$= \sideset{}{'}\sum_{n_1, n_2} \frac{1}{(n_1\omega_1 + n_2\omega_2)^{2k}} = G_k(\omega_1', \omega_2')\,.$$

We should like to reduce this function of two variables to a function of one variable. This we do by forming the ratio

$$\frac{G_2^3(\omega_1, \omega_2)}{G_3^2(\omega_1, \omega_2)} = \frac{\omega_2^{-12}G_2^3(\tau, 1)}{\omega_2^{-12}G_3^2(\tau, 1)} = \frac{G_2^3(\tau, 1)}{G_3^2(\tau, 1)}\,,$$

where we make use of the homogeneity of the G's. This is a function of τ which we shall call $\mathscr{F}(\tau)$. Now our long story about the rearrangement of the series yields the desired result, for from

$$\frac{G_2^3(\omega_1', \omega_2')}{G_3^2(\omega_1', \omega_2')} = \frac{G_2^3(\tau', 1)}{G_3^2(\tau', 1)} = \mathscr{F}(\tau')$$

we have:

$$\mathscr{F}(\tau') = \mathscr{F}(\tau),$$

where

$$\tau' = \frac{\omega_1'}{\omega_2'} = \frac{a\omega_1 + b\omega_2}{c\omega_1 + d\omega_2} = \frac{a\tau + b}{c\tau + d}.$$

Thus τ' is a modular transformation of τ as $\left|\begin{smallmatrix} a & b \\ c & d \end{smallmatrix}\right| = 1$. Hence

$$\mathscr{F}\left(\frac{a\tau + b}{c\tau + d}\right) = \mathscr{F}(\tau)$$

is a modular function.

After this long story we have succeeded in constructing a function that is invariant under the modular group, and we could show by a longer story, too long for these lectures, that $\mathscr{F}(\tau)$ is not constant.

Note to Chapter 10

The notion of a function belonging to a group which Rademacher introduces in this chapter is very closely allied to the concept of a function on a Riemann surface If one looks, for example, at transformations $z' = z + m_1 w_1 + m_2 w_2$, where w_1 and w_2 are two noncolinear complex numbers, then the fundamental region is a parallelogram with sides w_1 and w_2. This region can be realized topologically as the surface of a torus (doughnut) after glueing together the opposite (congruent) sides of this parallelogram. We have, therefore, realized the fundamental region as a Riemann surface, and the elliptic functions which Rademacher discusses are just functions on this Riemann surface.

Chapter 11

Linkages

We have talked at length about inversion and our discussion would be incomplete without a brief mention of the linkages which can be used to perform this operation. A linkage is a mechanism consisting of certain rods joined together by hinges, so that the system is free to move in a plane.

The first linkage to be described was invented by a French general named Peaucellier. In 1864 he communicated his discovery to the journal *Nouvelles Annales de Mathématiques*. However the editor suppressed this until after the same mechanism was rediscovered by the Russian Lipkin in 1871.

This mechanism, known as *Peaucellier's cell*, consists of six rods (Fig. 37). Four of these are of equal length, say c, and are joined together to form a rhombus by hinges at the vertices P, Q, R, and P'. The other two rods are also equal in length, say b, and join the rhombus by hinges at Q and R. These are joined together at a fixed point O, about which the whole mechanism is free to rotate in the plane.

In the figure we have tacitly assumed that $b > c$. P and P' are free to move closer together or further apart, within limits; however it is clear that O, P, and P' must lie on the same straight line, no matter how the machine is moved.

Now we wish to show that this linkage produces an inversion. Let us draw a few auxiliary lines (dotted in Fig. 37): a circle centered at Q of radius c which

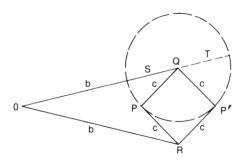

Figure 37

passes through P and P', and an extension of OQ. The line OQ extended is cut by the circle in two points S and T. Here we have the following geometrical situation: a circle cut by two secants (Fig. 38). In such a configuration $OA \cdot OB = OC \cdot OD$, for the triangles OAD and OCD are similar as they have a common angle at O and the angles at B and D are equal because they view the same segment of the circle, AC. Thus $OC/OA = OB/OD$, which is the assertion. From this it follows that $OP \cdot OP' = OS \cdot OT$. But we know from our construction that $OS = b - c$ and $OT = b + c$. Hence $OP \cdot OP' = b^2 - c^2$, a constant of the machine. Thus P and P' are inverse points with respect to the circle of radius $\sqrt{b^2 - c^2}$ centered at O.

This is indeed the circle of inversion of this linkage, for if we make the points P and P' coincide, then r is the third side of a right triangle of hypotenuse b and leg c (Fig. 39). This machine will invert points which lie inside the ring-shaped domain between the concentric circles about O of radii $b - c$ and $b + c$.

An interesting application of a property of inversion has been made using this apparatus. If the point P moves on a circle passing through O, then P' will move on a straight line that is parallel to the tangent to the circle at O (Fig. 40). To force P to move on such a circle, let us add a seventh rod to the system of length a which has one end in a hinge at P and the other in a fixed pivot at O', such that $OO' = a$.

This linkage is historically interesting as it provides a solution to the problem of finding a linkage which produces rectilinear motion. Engineers and mathematicians of the Nineteenth Century spent much time on this problem, and many first rate men failed to find a solution to it. The invention of the steam engine made

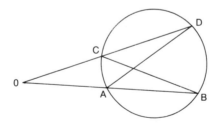

Figure 38

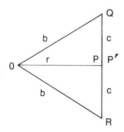

Figure 39

this a pressing problem, for it is highly desirable that the piston should move rectilinearly to maintain compression and avoid wear. The common solution, that of a crosshead moving on rails, as on the side of a locomotive, is not a linkage (Fig. 41). In an automobile engine the crosshead and the piston are combined into a unit, the skirt serving as the crosshead. The fact that this is only an approximation is clear to anyone familiar with old engines.

Watt's first steam engine contained a linkage, known as Watt's parallelogram, which is an approximate solution to the problem (Fig. 42). The piston is joined to one vertex of a hinged parallelogram; another vertex of the parallelogram is free; a third is joined by a hinged rod to a fixed pivot B; while the fourth vertex is connected by a rigid extension of the side CD to a second fixed pivot at A. If we draw a line from the vertex where the piston joins the parallelogram to the vertex A, which intersects the far side of the parallelogram in the point (*), from the similarity of the triangles so formed, we see that the role of the parallelogram is merely to magnify the motion of the point (*). Thus the essential parts of this device are the two pivots A and B and the three rods connecting them. Let us investigate the motion of the midpoint of the center rod, and for convenience let us assume that the center of coordinates lies midway between A and B, that the two outer rods are of the same length, say b, and that the central rod is of length c (Fig. 43). If the coordinates of the two hinges are (x_1, y_1) and (x_2, y_2), then M has the coordinates $\xi = (x_1 + x_2)/2$ and $\eta = (y_1 + y_2)/2$. If these are introduced

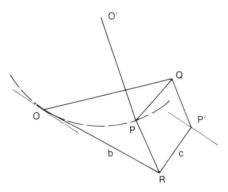

Figure 40

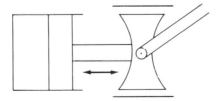

Figure 41

into the three equations $(x_1 + a)^2 + y_1^2 = b^2$, $(x_2 - a)^2 + y^2 = b^2$, and $(x_1 - x_2)^2 + (y_1 - y_2) = c^2$, with the constants adjusted so that $4b^2 = 4a^2 + c^2$, an equation of the sixth degree can be obtained for ξ and η;

$$(\xi^2 + \eta^2)^3 - 4a^2(\xi^2 + \eta^2 - a^2)\xi^2 - a^2c^2\eta^2 = 0.$$

The partial derivatives of the first and second order of this expression vanish at the origin which indicates that we have a double point (i.e., a point where the curve crosses itself) and a point of inflection for both branches of the curve. Thus the motion in a small region about the origin is approximately rectilinear. It is perhaps easier to see this fact geometrically (Fig. 44). Imagine that the arm pivoted at B is raised as high as it can go. At this moment the linkage forms a triangle with the x-axis. The arm at A still can be raised, and in doing so the arm at B must fall. At one instant the linkage will form a trapezoid with the x-axis. At this moment M will be at its highest point. The arm at B can be rotated downward until the position of the linkage is the reflection of the original position in the x-axis. Then continued motion repeats the same pattern in the lower half plane. M passes successively through the points 0, 1, 2, 3, and back to 0, so that it describes a very narrow curve somewhat like a figure-8, and in the neighborhood of the origin it is clear that the motion is approximately rectilinear.

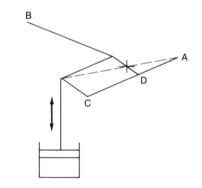

Figure 42

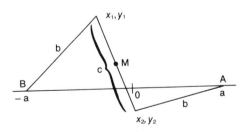

Figure 43

This problem has been very fascinating to many workers. Many great mathematicians of the Nineteenth Century attempted to solve the problem of producing
straight line motion by a linkage. One of the foremost of these was the Russian
Tchebysheff, who tried for many years to produce such a motion. As with all first-
rate mathematicians his unsuccessful efforts to solve this problem led to first-rate
new mathematics. It was through these attempts that he was led to invent the
polynomials that bear his name, which are polynomial approximations to a straight
line. He was almost convinced that the problem was insoluable. However Lipkin,
who did solve the problem, was one of his students. As is frequently the case
in mathematics, when a problem is solved once, other different solutions are
quickly found.

The Hart linkage which was discovered toward the end of the Nineteenth
Century solves the same problem as Peaucellier's cell and has the advantage that
it contains only four rods (see Fig. 45). Here we begin with a parallelogram hinged
at the corners, collapse it on itself, and then pull it up by two diagonal corners
to form an overlapping parallelogram.

However we move the parallelogram in this configuration (Fig. 46), it turns
out that the diagonals AC and BD remain parallel. This we can see by noting that
the triangles ABC and CDA must remain congruent and hence have equal altitudes.
Let us draw a straight line parallel to AC, defining the points O, P, P', and R where
it intersects the legs of the parallelogram.

These four points must remain colinear no matter how the figure is moved. First
let us keep the parallelogram in a fixed position. $AO:OD = AP:PB$ and

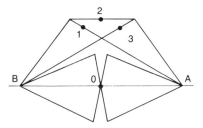

Figure 44

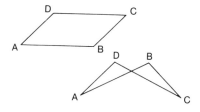

Figure 45

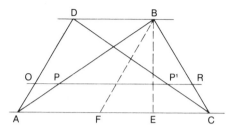

Figure 46

$CP':P'D = CR:RB$. Furthermore from the congruence of the triangles we have $AO:OD = CP':P'D$. These ratios will be invariant when the figure is moved. The first two mean that OP and $P'R$ are parallel to DB and hence to one another. The third means that OP' is always parallel to AC. But this in turn is parallel to DB; so we conclude that OP' and $P'R$ are parallel. But since they have a common point, they must be the same line. Hence O, P, P', and R remain colinear when the figure is moved.

Now we wish to show that P and P' are inverse points. From the similarity of the triangles we have:

$$OP:DB = AO:AD$$

and

$$OP':AC = DO:DA .$$

These we combine as

$$OP \cdot OP' = AC \cdot DB \frac{AO \cdot DO}{(AD)^2}$$

Now AO, DO, and AD are fixed constants determined by the dimensions of the linkage. However AC and DB are clearly variable. If P and P' are inverse points, then $AC \cdot DB$ must be constant. Let us drop a perpendicular from B with foot at E and reflect the line BC in this perpendicular to form the isosceles triangle FBC. Then we have easily that $AC = AE + EC$ and $DB = AE - EC$ so that $AC \cdot DB = AE^2 - EC^2$. Further $AE^2 + BE^2 = AB^2$ and $BE + EC^2 = BC^2$ so that $AE^2 - EC^2 = AB^2 - BC^2$, and the latter difference is a constant determined by the linkage. Hence $OP \cdot OP'$ is a constant and P and P' are inverse points with respect to a certain circle about O.

Thus if the point O is fixed as a pivot, this linkage is an inversor; and if we constrain the point P to move on a circle which passes through O, by adding a fifth arm about a new pivot O', then this linkage also produces rectilinear motion.

Rectilinear motion can also be produced by linkages without recourse to inversion. The double rhomboid of Kempe is such a machine (Fig. 47). It consists of two bars of equal length, say a, hinged together, say at A. Two other bars of equal length, say b, are hinged together and to the first pair to form a quadrilateral

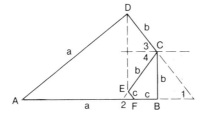

Figure 47

ABCD. Such a quadrilateral whose sides in adjacent pairs are of equal length we shall call a rhomboid. To complete the mechanism a second rhomboid similar to the first with one pair of sides equal to *b* and the other pair of length *c* determined by the ratio $a : b : : b : c$ is added to the figure so that *BC* is one of the sides of length *b*. Only two new bars, *CE* and *EF*, are needed to do this. The bar *AC* is rigid, although *EF* is hinged to it at the point *F* at a distance *c* from *B*.

These two rhomboids will always remain similar, no matter how the linkage is moved since they have proportional sides and the angle at *B* in common. If we extend *AB* until it intersects *DC* extended (dotted lines in the Fig. 47) to form the angle 1, and analogously for the smaller rhomboid extend *CE* to intersect *AB* (*BF* extended towards *A*) to form angle 2, from the similarity of the figures we have angle 1 = angle 2. Now drawing a dotted line through *C* parallel to *AB*, we form the angles 3 and 4. These are respectively equal to the angles 1 and 2, so that angle 3 equals angle 4. Thus the angle at *C* is bisected by the parallel to *AB*. Let us connect *D* and *E* by a dotted line. The triangle so formed is isoceles; hence the line *DE* is orthogonal to the parallel to *AB* through *C* and consequently orthogonal to *AB* itself. Thus if *AB* is constrained to move parallel to itself, *DE* will remain perpendicular to *AB*. Now if we make the point *D* fixed as a pivot and constrain *AB* to move parallel to itself, the point *E* will move in a straight line — the desired result. This constraint can be obtained by adding a new bar of length *a* from *B* to a fixed pivot *G* such that *ABGD* is a parallelogram. Then as the linkage is moved, *AB* will remain parallel to itself and *E* will describe a straight line.

The Kempe rhomboid can be used to do a little more. Suppose we have two copies of it joined nose to nose at *C* as in Fig. 48. Now we modify the linkage

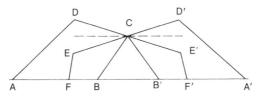

Figure 48

so that DCE' and ECD' are rigid bars pivoted at C. When this linkage is moved, the two rhomboids will remain similar as the two angles at C will be equal. From our previous argument it is clear that AB and $B'A'$ must be parallel as they are both parallel to the bisector of the common angle at C. However we should like to show that these two lines are in fact the same line. This would follow if we could show that the two rhomboids were congruent.

A rhomboid is determined if we know the sides and an angle between two adjacent sides. Here we know the sides, but only the angle at C which is common to the two rhomboids. This is enough to determine the rhomboid; in fact we need only the half angle common to both. Suppose the problem of construction were solved. Then if we reflect the point C into H, we see that the angle 1, which we know to be equal to the angle 2, is in turn equal to the angle 3 (Fig. 49). Thus knowing the angle 1 and the sides of the rhomboid, we can determine H and hence the whole rhomboid. Since the sides of our two rhomboids are by supposition equal and the angle at C is common to both, the foregoing shows that they are congruent. Hence AB is indeed colinear with $B'A'$. Now the utility of the two double rhomboids is clear. If we clamp down AB, then $B'A'$ will be constrained to move along the extension of AB. With our previous linkages we have only been able to move points in a straight line; now we have a device to which we can clamp a portion of the plane and shift the plane parallel to itself.

The foregoing linkages do produce rectilinear motion, but for the engineer they would not be a complete solution to the problem, for additional constraints are necessary to keep the parts of the linkages in the plane. A practical linkage would be one that produced rectilinear motion in space. Such a linkage was actually found by Sarrus in 1853 before the solutions that we have just described.

This linkage is formed from plates that are hinged together. One way to form such a space linkage would be to add a third dimension to the rods of our plane linkages, thus making prism-like linkages. However, we can do this in a different way, by making a pyramidal linkage. Imagine a sphere on the surface of which we have a quadrilateral, the vertices of which are connected to the center (Fig. 50). This forms a pyramid. We now imagine the sides of this pyramid as being plates and the edges of the pyramid as being hinges. This we call a pyramidal linkage. It can be deformed into various shapes. Now suppose we have two such pyramidal linkages which have one edge in common, as in Fig. 51. This common hinge

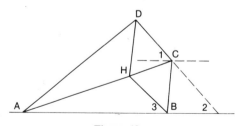

Figure 49

can be removed from the linkage without affecting the performance of the machine — the plates A and B will continue to behave as if they were rotating about this hinge even if it is not present. This notion of a non-material hinge is actually used in some practical machines — it is not an abstraction of mathematics.

Now if this hinge doesn't need to be materially present, what happens if we remove it to infinity? Then A and B become parallel plates and they rotate about the line at infinity which corresponds to the intersection of two such parallel planes (Fig. 52). The hinges a, b, c, having a common point on this line which has been removed to infinity, must become lines parallel to the planes A and B in a certain direction. Similarly d, e, f go over into parallel lines in another direction.

Since a hinge resists motion except in a direction around its axis and since the directions of the two sets of hinges are different, it is clear that this linkage permits motion only in a direction perpendicular to the planes A and B.

This linkage of Sarrus — a polyhedral linkage — can be materialized by slitting the edges of a closed box and carefully creasing the sides.

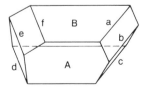

Figure 50

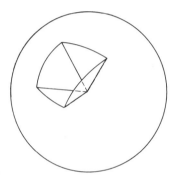

Figure 51

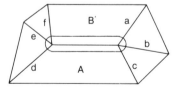

Figure 52

Bibliography

Maxime Böcher, *Introduction to Higher Algebra*, Macmillan, New York, 1949.

G. Birkhoff, *A Source Book in Classical Analysis*, Harvard University Press, Cambridge, Mass., 1973, pp. 98–103.

E. Bombieri, Le grand crible dans la théorie analytiques des nombres, Asterisque, **18** (1974).

J. Chen, On the representation of a large even integer as a sum of a prime and the product of at most two primes, Sci. Sinica, **16** (1973), 157–176.

L. Dirichlet, *Mathematische Werke*, Chelsea, New York, 1969.

F. J. Dyson, The approximation to algebraic numbers by rationals, Acta Math. Acad. Sci. Hungar. **79** (1947), 225–240.

Euclid, *The Thirteen Books of Euclid's Elements*, Dover, New York, 1956.

L. R. Ford, Fractions, Am. Math. Monthly, **45** (1938), 586–601.

A. O. Gelfond, The approximation of algebraic numbers by algebraic numbers and the theory of transcendental numbers, American Mathematical Society Translation No. 65 (1952).

H. Halberstam, H. E. Richert, *Sieve Methods*, Academic, London, 1974.

G. H. Hardy, *Collected Papers of G. H. Hardy*, Clarendon, 1966.

C. Hooley, *Applications of Sieve Methods to the Theory of Numbers*, Cambridge University Press, Cambridge, England, 1976.

A. Hurwitz, *Mathematische Werke von Adolf Hurwitz*, Vol. II, Birkhäuser Verlag, Basel-Stuttgart, 1963, 122–128.

M. N. Huxley, On the difference between consecutive primes, Inventiones Math, **15** (1972), 164–170.

H. Iwaniec, M. Jutila, Primes in short intervals, Arkiv Matematik, **17** (1979), 167–176.

E. Landau, *Elementary Number Theory*, Chelsea, New York, 1958.

D. N. Lehmer, *List of Prime Numbers from 1 to 10,006,721*, Hafner, New York, 1956.

J. Liouville, Sur des classes très-étendues de quantités dont la valeur n'est ni algébrique, ni même réductible à des irrationnelles algébriques, J. Math. Appl., **16** (1851).

H. Montgomery, Topics in multiplicative number theory, Lecture Notes in Mathematics 227, Springer–Verlag, Berlin, 1971, 130–132.

H. Montgomery, Zeros of L-functions, Inventiones Math., **8** (1969), 346–354.

G. Polya, G. Szego, *Aufgaben und Lehrsätze aus der Analysis,* Vol. II [VII, 21] Springer–Verlag, Berlin, 1964, 119.

B. Riemann, *Gesammelte Mathematische Werke,* Teubner, Stuttgart, 1978, 145–155.

K. F. Roth, Rational approximations to algebraic numbers, Mathematika, **2** (1955), 1–20.

C. L. Siegel, Approximation algebraischer Zahlen, Math. Z., **10** (1921).

A. Thué, Über Annäherungswerte algebraischer Zahlen, J. Reine Angew. Mathem., **135** (1909).

J. V. Uspensky, *Introduction to Mathematical Probability,* McGraw-Hill, New York, London, 1937.

I. M. Vinogradov, Representation of an odd number as a sum of three primes, Dokl. Akad. Nauk. USSR, **15** (1937), 169–172.

Index